Lecture Notes in Electrical Engineering 228

For further volumes:
http://www.springer.com/series/7818

Yuliang Zheng

Tunable Multiband Ferroelectric Devices for Reconfigurable RF-Frontends

Author
Dr. Yuliang Zheng
TU Darmstadt
Darmstadt
Germany

ISSN 1876-1100 e-ISSN 1876-1119
ISBN 978-3-642-35779-4 e-ISBN 978-3-642-35780-0
DOI 10.1007/978-3-642-35780-0
Springer Heidelberg New York Dordrecht London

Library of Congress Control Number: 2012954618

© Springer-Verlag Berlin Heidelberg 2013

This work is subject to copyright. All rights are reserved by the Publisher, whether the whole or part of the material is concerned, specifically the rights of translation, reprinting, reuse of illustrations, recitation, broadcasting, reproduction on microfilms or in any other physical way, and transmission or information storage and retrieval, electronic adaptation, computer software, or by similar or dissimilar methodology now known or hereafter developed. Exempted from this legal reservation are brief excerpts in connection with reviews or scholarly analysis or material supplied specifically for the purpose of being entered and executed on a computer system, for exclusive use by the purchaser of the work. Duplication of this publication or parts thereof is permitted only under the provisions of the Copyright Law of the Publisher's location, in its current version, and permission for use must always be obtained from Springer. Permissions for use may be obtained through RightsLink at the Copyright Clearance Center. Violations are liable to prosecution under the respective Copyright Law.

The use of general descriptive names, registered names, trademarks, service marks, etc. in this publication does not imply, even in the absence of a specific statement, that such names are exempt from the relevant protective laws and regulations and therefore free for general use.

While the advice and information in this book are believed to be true and accurate at the date of publication, neither the authors nor the editors nor the publisher can accept any legal responsibility for any errors or omissions that may be made. The publisher makes no warranty, express or implied, with respect to the material contained herein.

Printed on acid-free paper

Springer is part of Springer Science+Business Media (www.springer.com)

Preface

This book is a summary and a precious memory of my research in the Institute of Microwave Engineering and Optics in the Technische Universität Darmstadt.

First of all I would like to express my gratitude to my supervisor Prof. Dr.-Ing. Rolf Jakoby. Without his profound advise and warm encouragement, this work would never have become a reality. My sincere appreciation goes also to Prof. Dr. Andreas Klein for the great cooperation and his helpful guidance throughout the years.

For the inspiring discussions and constructive cooperation, my fellow colleagues in the Institute of Microwave Engineering and Optics are warmly thanked. Especially, I deeply appreciate Dr.-Ing. Patrick Scheele, Dr.-Ing. Andre Giere, Mohsen Sazegar, Dr.-Ing. Xianghui Zhou, Alexander Gäbler, Onur Hamza Karabey and Dr.-Ing. Holger Maune for the help with no reservation and the cordial friendship, Arshad Mehmood, Mohammad Nikfal Azar, Erick González Rodríguez and Alex Wiens for the fruitful teamwork. Without them, the endeavor would be a lonely journey.

The interdisciplinary cooperations with the partners from the Department of Materials Science in Technische Universität Darmstadt, the Institute of Microelectronic Systems in Technische Universität Darmstadt, the Institute for Materials Research in Karlsruhe Institute of Technology, the Department of Microsystems Engineering in University of Freiburg, and EPCOS AG have enabled the work. For bringing me a new horizon I would like to acknowledge Prof. Dr.-Ing. Jürgen Haußelt, Dr. Joachim R. Binder, Prof. Dr. Lambert Alff, Prof. Dr. Dr. h.c. mult. Manfred Glesner, Dr. Robert Schafranek, Dr. Shunyi Li, Karsten Rachut, Dr. Philipp Komissinskiy, Stefan Hirsch, Dr.-Ing. Ping Zhao, Dr. Florian Paul, Henrik Elsenheimer, Dr. Holger Gesswein, Dr. Stefan Seitz, and Dr. Anton Leidl.

Finally I owe my greatest debts to my family. I thank my beloved parents for life and the unfailing encouragement. I thank my sons, who remind me that the exploration is a beauty of the life. Special thanks to my wife Dr. phil. Ying Cai for being a fantastic partner and wonderful mother.

Abstract

This book focuses on the enabling ferroelectric device technology in reconfigurable RF frontends for frequency-agile, software-defined and cognitive radios, which are expected to cope with the demands of multiband and multi-standard operation. The ferroelectric devices are an emerging technology based on the nonlinear solid state ferroelectric materials. Such materials e.g. Barium-Strontium-Titanate $Ba_xSr_{1-x}TiO_3$, or namely BST, exhibit a variable permittivity dependent on external electric field. The high permittivity, adequate tunability and dielectric loss as well as low leakage current allow the implementation in compact passive tunable devices. BST can also be realized in several technologies such as ceramic thick-film and crystal thin-film, which can be flexibly adapted to applications.

The investigation ranges from micro processing technologies to the development of novel devices. Within interdisciplinary cooperations, the materials are synthesized, processed and characterized. Micro processing technologies for device realization are revised. Several novel devices are addressed. By introducing thin conductive layer in the middle of two stacked BST thin-films, the unavoidable generation of acoustic resonance due to the electrostriction is considerably suppressed. When a highly resistive alumina oxide film is used to engineer the band gap on the BST surface, the resulted controllable electron injection leads to the innovative programmable bi-stable high frequency capacitors. The kernel frontend parts, including multiband tunable antenna, tunable matching network and tunable substrate integrated waveguide filter are addressed at the end. The challenging trend of antennas towards compactness with wider spectrum coverage is coped with several tunable resonant modes of the antennas with integrated BST varactors. A prototype optimized for frequency division duplex services covers 1.47 GHz to 1.76 GHz with a variable distance from 38 MHz to 181 MHz between up- and down-link channels. The environmental impact and frequency dependence of antennas can be compensated by tunable matching networks. This work demonstrates an integrated ferroelectric thick-film tunable matching network covering 1.8 GHz to 2.1 GHz. It exhibits an insertion loss between 0.86 dB and 0.98 dB, in a 3 mm × 3 mm multilayer package. At last a compact bandpass filter is enabled by an evanescent mode substrate integrated waveguide cavity in ferroelectric ceramics, integrated with tunable complementary split ring resonator scatterers and tunable matching networks. A compact demonstrator covers 2.95 GHz to 3.57 GHz with a 3 dB fractional bandwidth of up to 5.4 % and a comparatively low insertion loss between 3.3 dB and 2.6 dB.

Contents

1	**Introduction**	1
2	**Tunable Microwave Dielectrics**	3
	2.1 Dielectric Properties	3
	2.2 Pyroelectricity	5
	2.3 Ferroelectricity	6
	2.4 Piezoelectricity and Electrostriction	9
	2.5 Electrostatic Properties	11
3	**Processing of Ferroelectric Films and Components**	13
	3.1 Processing of BST Thick-Film Components	13
	3.1.1 Screen-Printed Thick-Film Ceramics	13
	3.1.2 Planar Circuitry Realization and Reliability	17
	3.2 Processing of BST Thin-Film Components	25
	3.2.1 BST Thin-Film by RF Sputtering and Pulsed Laser Deposition	25
	3.2.2 Realization of Thin-Film Components	28
	3.3 Characterization and Modeling of Lossy Varactors	31
4	**Novel Multilayer Components on BST Thin-Films**	37
	4.1 Acoustic-Free BST Thin-Film Varactor	37
	4.1.1 Acoustic Properties	38
	4.1.2 Modeling of Multilayer Acoustic Resonator	43
	4.1.3 Suppression of Acoustic Resonances	46
	4.2 Programmable Bi-stable Capacitor	49
	4.2.1 Non-volatile Bi-stability by Induced Hysteresis	50
	4.2.2 Long-Term Stability	54
5	**Tunable Multiband Ferroelectric Devices**	55
	5.1 Tunable Multiband Antennas	56
	5.1.1 Theoretical Analysis: Q Factor, Bandwidth and Tunability	57

		5.1.2	Nonlinear Model of Capacitively Loaded Planar Inverted-F Antenna.................................	66
		5.1.3	Suppression of Harmonic Radiation by BST Thick-Film Varactors	76
		5.1.4	Independently Tunable Multiband Slot Antenna..........	79
		5.1.5	Integrated Duplex Ceramic Antenna	84
	5.2	Tunable Single-Band Impedance Matching Network for Antennas...		87
		5.2.1	Theoretical Analysis: Single-Band Design Guidelines.....	88
		5.2.2	Optimal Design with Lossy Components	97
		5.2.3	Antenna Bandwidth Enhancement	100
		5.2.4	System in Package Realization	104
		5.2.5	Efficiency of Adaptive Control	110
	5.3	Tunable Multiband Matching Network		116
		5.3.1	Polynomial Synthesis Method for Impedance Matching ...	116
		5.3.2	Polynomial Optimization for Component Loss Compensation	121
		5.3.3	Impedance Matching Ranges	124
		5.3.4	Proof of Concept	127
	5.4	Tunable Substrate Integrated Waveguide Bandpass Filter		130
		5.4.1	Bandpass Filter Design	131
		5.4.2	Tunable Evanescent-Mode Substrate Integrated Waveguide Cavity..................................	132
		5.4.3	Tunable Impedance Matching Network	133
		5.4.4	Realization and Measurement	135

6 Summary and Outlook 137

A Technology Parameters 141

References.. 143

Glossary

BGA	= ball grid array.
C-V	= capacitance-voltage.
CPW	= coplanar waveguide.
CSRR	= complementary split-ring resonator.
DAC	= digital to analog converter.
DCS	= digital cellular system.
FBAR	= film bulk acoustic wave resonator.
FDD	= frequency division duplex.
FDTD	= finite-difference time-domain method.
FEM	= finite element method.
GSG	= ground-signal-ground.
GSM	= global system for mobile communications.
HTCC	= high temperature cofired ceramics.
IDC	= interdigital capacitor.
ITO	= indium tin oxide.
LMS	= least mean squares.
LTCC	= low temperature cofired ceramics.
MEMS	= microelectromechanical systems.
MIM	= metal-insulator-metal.
MOS	= metal-oxide-semiconductor.
PCS	= personal communications service.
PIFA	= planar inverted-F antenna.
PLD	= pulsed laser deposition.
PVD	= physical vapor deposition.
RF	= radio frequency.
RFID	= radio frequency identification.
RFT	= real frequency technique.
RHEED	= reflection high-energy electron diffraction.
SAW	= surface acoustic wave resonator.
SDR	= software-defined radio.

SIP	= system-in-package.
SIW	= substrate integrated waveguide.
SMA	= subminiature version a.
SoC	= system on chip.
SRR	= split-ring resonator.
TMN	= tunable matching network.
TPG	= transducer power gain.
TRP	= total radiated power.
UBM	= under bump material.
UMTS	= universal mobile telecommunications system.
VNA	= vector network analyzer.

Chapter 1
Introduction

In last few decades the development of wireless technology has enabled the exciting growth in mobile telecommunications. The endless pursuit of wider bandwidth, higher frequency and better efficiency has driven the innovations of micro- and millimeter wave systems and devices, where the technologies that balances both performance and feasibility prevail. Nowadays the challenge of multiband and multi-standard operation is expected to be met by the frequency-agile, software-defined and cognitive radios. Similar to the reconfigurability of modulation scheme, time division, channel equalization and protocols in the processing unit at the base-band, through a dynamical reconfiguration of the high frequency frontend, the system's operation frequency, bandwidth, power level, as well as spatial diversity can be adapted to cope with the time and regional variations of traffic demands, while a high efficiency can be stabilized under changing environment. As an enabling technology, the underlying tunable devices have been attracting increasing research efforts. Various competitive tunable technologies have been proposed e.g. semiconductors, microelectromechanical systems, piezoelectric actuators and ferroelectric devices. The ferroelectric devices are an emerging technology which grounds on the nonlinear solid state ferroelectric materials. Such materials exhibit a variable permittivity depending on external electric field. Among the various ferroelectric material systems, the Barium-Strontium-Titanate $Ba_xSr_{1-x}TiO_3$, or namely BST, has shown a high permittivity, adequate tunability and dielectric loss, and low leakage current, which are indispensable for compact passive tunable devices. In the meantime, BST can be realized in several technologies like ceramic thick-film and crystal thin-film, which can be flexibly adapted to applications. The tunable devices generally consist of the ferroelectric films, metallic circuitries and additional functional layers. The innovation of such devices encompasses not only circuitry concepts but also fundamental materials. In the meantime the integration and control techniques are of importance as well.

The focus of this work ranges from the investigation of micro processing technologies to the development of novel components. Chapter 2 introduces the fundamental properties of ferroelectric materials. In chapter 3, the technologies for material deposition and component realization are reviewed. Novel processing methods are proposed, e.g. vertical connections through BST films and capillary formed by high temperature sacrificial layers. Component reliability is also investigated. Then electric properties are extracted and modeled.

The reconfigurability of frontends roots in the fundamental components like varactors. Chapter 4 addresses novel components built on BST thin-film. The multilayer thin-film technology is extended to accommodate additional layers. When a 20 nm thin conductive layer is introduced in the middle of two stacked BST films while antipolarized electric fields are applied to the two layers of BST film, the unavoidable generation of acoustic resonance due to the electrostrition is considerably suppressed. The method shows the potential to widen the applicable frequency range of BST thin-film varactors. Furthermore, a thin highly resistive aluminum oxide layer is inserted at the interface between metallic electrode and BST thin-film. It introduces a controllable electron injection, which allows a charge storage at the interface between the alumina layer and the BST film. The stored charge biases the film without external field. Hence the capacitor can switch between two states, and hold it. With the induced hysteresis, low loss capacitors are achievable in contrast to the capacitors in ferroelectric phase, since the thin-film is still in paraelectric phase. Hence, an innovative bi-stable high frequency capacitor is proposed for the first time.

The tunable ferroelectric devices including multiband tunable antenna, tunable matching network and tunable substrate integrated waveguide filter are addressed in chapter 5. The challenging trend of antennas towards compactness with wider spectrum coverage is coped with several tunable resonant modes. Such tunable antennas are integrated with BST varactors. On one hand, they provide narrower stationary bandwidth than the static counterparts. On the other hand, they can be tuned to the desired frequencies, which equivalently increase the spectrum coverage and allow multiband operation. The environmental impact and frequency dependence of antennas can be compensated by the tunable matching network based on BST varactors. Through an efficient control, it can not only stabilize the antenna impedance, maintain the transducer gain, but also increase the frequency coverage. The complete design methodology for single band matching networks is proposed, and afterwards extended to multiband designs. The design methods also consider the varactor loss at the circuit and transmission polynomial function levels, which help to efficiently reduce the insertion loss. Aiming at a bandpass filter, an evanescent mode substrate integrated waveguide cavity in ferroelectric ceramics is proposed. It is loaded with a pair of complementary split ring resonators, which are tuned through embedded varactors. The input mismatch due to impedance drift during tuning is compensated by integrated tunable matching networks with varactors. The above investigations generally root in the analytic modeling and full-wave simulations. Therefore, the theoretical performance boundaries are identified, and the requirements on ferroelectrics' tunability and loss are recognized. All the concepts are finally proved by demonstrative prototypes.

A summary of the investigation comes at the end. An outlook of further development is also given.

Chapter 2
Tunable Microwave Dielectrics

Ferroelectric materials, or ferroelectrics for short, are usually complex oxide dielectrics with multiple functionalities. Their physical properties including polarization, permittivity etc. are generally susceptible to changes of temperature, mechanical strain and external electromagnetic field. Specifically, the ferroelectrics with a permittivity dependent on external electric field are a subset of the thermal sensitive pyroelectrics, which are in turn a part of the piezoelectrics with electrical response to applied mechanical field [52]. These inherent multiple functionalities have been implemented in various applications. Pioneer work on ferroelectricity was focused on the Rochelle salt since 1920s [26]. In the middle of 1960s, ferroelectrics came into the view of tunable microwave applications [22]. Later in 1970s, the demands toward nonvolatile memory triggered the development of ferroelectric film technologies. In last decades, research efforts focus on the integration of ferroelectric films onto silicon and ceramic substrates.

Two classes of ferroelectrics have been found, namely the order-disorder type and displacive type. Below the phase transition temperature, the ferroelectricity of former type relates to the ordering of the ions, as in the hydrogen bounded KH_2PO_4. For displacive type, the spontaneous polarization is formed by the displacement of crystal sublattices in relation to others. In complex oxide dielectrics, e.g. perovskite crystals, the ferroelectricity is in general the displacive type. In the following, the exemplary perovskite system $Ba_xSr_{1-x}TiO_3$ is considered.

This chapter introduces the essential properties of the $Ba_xSr_{1-x}TiO_3$ system. It provides a groundwork for the discussion in later chapters on realization technologies, component and device conceptions. The nonlinear dielectricity, temperature dependent phase transition, acoustic properties and leakage current are to be addressed.

2.1 Dielectric Properties

Dielectric materials are electrical insulators. In an external electric field, they show no flowing of electric charges, but a shift from their average equilibrium positions, which results in dielectric polarization. When the positive charge displaces along the external field and the negative charge shifts towards the opposite, an internal electric field establishes and partially compensates the external one inside dielectric [11]. In the simplified scenario of two point charges, the electric dipole moment $\vec{p}_{dip}$ is determined by the displacement $\vec{d}$ of the bounded charge pointing from the negative charge to the positive one, and the point charge q:

$$\vec{p}_{dip} = q \cdot \vec{d} . \tag{2.1}$$

When a bulk material is polarized in an inhomogeneous external electric field, the dipole moment density of the continuously distributed charge in a confined volume V is therefore determined by the local density of the bounded charge ρ:

$$\vec{p}_{dip}(\vec{r}_0) = \oint_V \rho(\vec{r})(\vec{r} - \vec{r}_0) dv(\vec{r}) . \tag{2.2}$$

In the ferroelectric systems, the dipole moment is dominant in the high frequency electromagnetic scattering behavior, in comparison with the higher quadrapole moment. The dipole moment density is sufficiently accurate to represent the polarization density $\vec{P}$ in Maxwell's equations:

$$\vec{P}(\vec{r}_0) = \vec{p}_{dip}(\vec{r}_0) . \tag{2.3}$$

Maxwell's equations extend the Gauss's law by introducing the electric displacement field $\vec{D}$, and relate the $\vec{D}$ and $\vec{P}$ with free charges and bounded charges:

$$\vec{D} = \varepsilon_0 \vec{E} + \vec{P} . \tag{2.4}$$

In reference to the material's permittivity, the intrinsic polarization equivalently alters it. Therefore the susceptibility χ_e is defined as the proportionality relating the electric field $\vec{E}$ to the dielectric polarization density $\vec{P}$:

$$\chi_e(\vec{E}) = \frac{1}{\varepsilon_0} \frac{\partial \vec{P}(\vec{E})}{\partial \vec{E}} . \tag{2.5}$$

A linear permittivity of a homogeneous material is then related to that of free space, by the relative permittivity ε_r:

$$\begin{aligned} \vec{D} &= \varepsilon_0 \vec{E} + \varepsilon_0 \chi_e \vec{E} \\ &= \varepsilon_0 (1 + \chi_e) \vec{E} \\ &= \varepsilon_0 \varepsilon_r \vec{E} , \end{aligned} \tag{2.6}$$

where

$$\varepsilon_r = (1 + \chi_e) . \tag{2.7}$$

In response to alternating field, the dielectrics exhibits a complex permittivity. It is separated in real and imaginary parts as:

$$\varepsilon(\omega) = \varepsilon'(\omega) + j\varepsilon''(\omega) \,. \tag{2.8}$$

The real part $\varepsilon'(\omega)$ refers to the energy storage in the material, while the imaginary part $\varepsilon''(\omega)$ refers to the energy dissipation. To evaluate generally the loss behavior, a loss tangent $\tan \delta$ is defined as the quotient between the imaginary part and the real part:

$$\tan \delta = \frac{\varepsilon''(\omega)}{\varepsilon'(\omega)} \,. \tag{2.9}$$

It is actually the tangent of the angle between the resistive loss component and the reactive storage component on a complex plane.

2.2 Pyroelectricity

Certain dielectric materials show a temporary voltage when their temperature varies. This property is named pyroelectricity. There, the change in temperature alters the positions of the atoms slightly within their crystal structures, and therefore, the polarization of e.g. the dipole moment changes, which leads to a voltage across the crystal. In the meantime, the permittivity ε_r changes as shown in Eqn. 2.5 and 2.7. If the temperature anchors, the generated voltage reduces due to the intrinsic leakage current through the crystal. However, the permittivity change maintains. If there is no phase transition, the temperature dependence of permittivity can be modeled by Curie-Weiss law [37]. The permittivity ε is determined by both the temperature T and a Curie constant C, whereas the T_c denotes the characteristic Curie temperature:

$$\varepsilon = \frac{C}{T - T_c} \,. \tag{2.10}$$

As an example, BST ($Ba_xSr_{1-x}TiO_3$) is a mixed system of Strontium Titanate ($SrTiO_3$) and Barium Titanate ($BaTiO_3$). By varying the proportion between Strontium and Barium ingredient, the pyroelectricity of BST can be tailored between that of $BaTiO_3$ and $SrTiO_3$ [48, 104]. As shown in Fig. 2.1(a), the permittivity shows strong dependencies on two determining factors [48]: first, for the given $Ba_{0.6}Sr_{0.4}TiO_3$ with the ratio from Ba to Sr of 6 : 4, the permittivity shows a maximum at -2 °C. When either increasing or decreasing the temperature, the permittivity drops drastically to about 10 % of the maximum. Second, when changing the Ba to Sr ratio, the temperature dependence alters accordingly.

In the meantime, when the dimension of the material reduces, e.g. to sub-micrometer and nanometer range, the change of crystallization phases of the material imposes additional influence on the temperature dependence. As illustrated in Fig. 2.1(b), with a same stoichiometry of $Ba_{0.7}Sr_{0.3}TiO_3$, the permittivity of BST thin-film differs considerably from that of a BST bulk ceramic. Not only the amplitude but also the temperature of the maximal permittivity alters. Furthermore, the sharpness of the temperature dependence reduces drastically as well [86]. To explain this, a model of serial capacitors can be employed. On the surfaces of a BST thin-film, there are depleted layers or namely dead layers. In the dead layer,

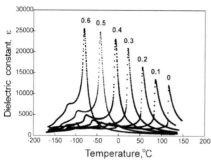 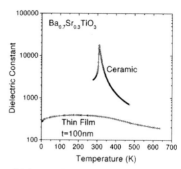

(a) Temperature dependency of permittivity, and the influence of Sr ingredient

(b) Comparison of the temperature dependency between BST bulk ceramic and thin film.

Fig. 2.1 The pyroelectricity of $Ba_xSr_{1-x}TiO_3$ in dependency of Sr ingredient [48], and a comparison between $Ba_{0.7}Sr_{0.3}TiO_3$ bulk ceramic and 100 nm-thin film which have the same stoichiometry [86]

the permittivity reduces considerably [4, 89, 95]. Such layer appears not only between the film, electrodes and substrate, but also between the internal crystal grains [57, 90]. Considering the dimension of the thin-films, which is close to the diameter of crystal grains, e.g. several hundred nanometers, the dead layer at the grains boundaries can be more influential than that of bulk ceramics. The source of such layers is still controversial. Possible cause can be external mechanisms, e.g. the barrier between the substrate and the film [18, 74, 101], defects of the film on the surface [95]. Besides, several theories also address the internal causes, e.g. dipole-dipole-interaction on the surface [67, 116] and the screening effect of electrodes [10, 89, 105].

For microwave applications, such temperature dependence can on one hand reduce the applicable temperature range of the components based on the dielectric materials. On the other hand, it leads to innovative temperature sensitive components [64], utilizing the permittivity to modulate either phase or resonant frequency as sensing parameters.

2.3 Ferroelectricity

Certain pyroelectric materials possess a spontaneous electric polarization, which can be altered with an external electric field. This property is termed as ferroelectricity. The physical reason for ferroelectricity is the ionic motion in the crystals. The vibration of the ions, or namely lattice wave, couples with the external electromagnetic wave. Following the definitions in optics, one can factor the coupling into two perpendicular phonon components, namely the longitudinal optical phonon LO and transverse optical phonon TO. LO refers to the condition that the deflection direction **u** of the external wave is parallel with wave vector **q**. For TO, **u** is perpendicular to **q** [52]. In the exemplary BST system, there are three different TOs in the THz to infrared frequency range as illustrated in Fig. 2.2. They resonate at different frequencies.

2.3 Ferroelectricity

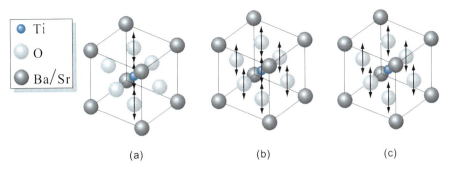

Fig. 2.2 Illustration of the transverse optical phonons in BST. (a) $TO\,1$: vibration of the linear chain of O^{2-}-Ti^{4+}-O^{2-} at about 16.2 THz. (b) $TO\,2$: vibration of Ti^{4+}- and O^{2-}-ions in relation to the Sr/Ba lattice at about 5.4 THz. (c) $TO\,3$: vibration of the oxygen octahedron in relation to all positive ions at about 2.61 THz [108].

In particular, $TO\,3$ resonates at the lowest frequency, which is essential to explain the ferroelectricity in microwave frequency range. In the harmonic resonators formed by vibrating positive and negative ions, the restoring force relates to the resonant frequency. The oscillator exhibits a lower restoring force with a lower resonant frequency, i.e. the phonon mode is softer than the higher frequency modes. When an external electric field is applied, the restoring force of $TO\,3$ is strengthened while the resonant frequency increases. Following the Lyddane-Sachs-Teller-relation, the resonate frequencies of TO and LO are related to the static and optical relative permittivity namely $\varepsilon_{r,s}$ and $\varepsilon_{r,\infty}$, respectively [61]:

$$\frac{\varepsilon_{r,s}}{\varepsilon_{r,\infty}} = \frac{\omega_{LO}^2}{\omega_{TO}^2}. \qquad (2.11)$$

The relation shows that, when the resonant frequency of $TO\,3$ increases, the static permittivity decreases. Since the microwave frequency is much lower than the resonant frequency of $TO\,3$, the permittivity can be approximated by the static permittivity. This explains the field dependence of the permittivity at microwave frequency. Furthermore, the freezing of the soft mode below the Curie temperature leads to the ferroelectric phase transition. Above the Curie point, there is no spontaneous polarization.

The ferroelectricity can be further modeled through thermodynamic theory. The free energy relationship is valid for both paraelectric and ferroelectric phases [37]:

$$F(P,T) = \frac{1}{2}\alpha P^2 + \frac{1}{4}\beta P^4, \qquad (2.12)$$

where P is the polarization density. Here only even terms are considered, since the free energy of ferroelectrics does not depend on the external field direction.

The electric field strength E is the derivative of the free energy as:

$$E = \frac{\partial F(P,T)}{\partial P} = \alpha P + \beta P^3 , \qquad (2.13)$$

and

$$\alpha = \frac{1}{\varepsilon_r \varepsilon_0} , \qquad (2.14)$$

where the higher order terms are omitted. It can be seen that the first order coefficient α represents the inverse dielectric permittivity.

If applying the Curie-Weiss law in Eqn. 2.10, the coefficient turns to be dependent on temperature:

$$\alpha = \frac{T - T_c}{C\varepsilon_0} , \qquad (2.15)$$

where only the transition between paraelectric and ferroelectric phases are considered. Hence the phase transition temperature equals to Curie-Weiss temperature T_c.

At T_c the spontaneous polarization P_s is the internal polarization when the external electric field strength $E = 0$. The polarization can be determined by:

$$\alpha P_s + \beta P_s^3 = E = 0 . \qquad (2.16)$$

The second root of P_s is given by [37]:

$$P_s = \sqrt{\frac{T_c - T}{\beta \varepsilon_0 C}} . \qquad (2.17)$$

At the Curie temperature where $T = T_c$, P_s is zero. Therefore, it is the phase transition temperature. Below T_c, the material is in ferroelectric phase, while above it, the material is in paraelectric phase, where there is no spontaneous polarization. By taking Eqn. 2.14, the permittivity can be calculated as:

$$\varepsilon_r = \frac{\partial E}{\varepsilon_0 \partial P} = \frac{1}{\varepsilon_0 (\alpha + 3\beta P^2)} = \frac{C}{T - T_c + 3C\varepsilon_0 \beta (\chi_e E)^2} . \qquad (2.18)$$

Since the susceptibility χ_e also depends on the external electric field, the relationship between permittivity, temperature and electric field strength turns to be implicit. However, it can be seen that the permittivity is altered by both temperature and electric field strength. Specifically, when there is no external electric field, the permittivity follows exactly the Curie-Weiss law as in Eqn. 2.10.

If the material is used as the dielectric layer in a parallel-plate capacitor, the capacitance can be simply modeled as:

$$C = \varepsilon_0 \varepsilon_r \frac{A}{d} , \qquad (2.19)$$

where A is the area of the planar electrode, d is the thickness of the ferroelectric layer. Here the fringe effect is neglected.

2.4 Piezoelectricity and Electrostriction

When the permittivity changes under certain electric field strength E, the capacitance changes accordingly. If compared to the properties without any external field, the relative change of the permittivity and the capacitance can be related as following:

$$\tau \equiv \frac{\varepsilon(0) - \varepsilon(E)}{\varepsilon(0)} = \frac{C(0) - C(V)}{C(0)}, \qquad (2.20)$$

where $C(0)$ is the capacitance without external voltage, $C(V)$ is the capacitance with a voltage V between electrodes. For engineering purposes, a more referential performance indicator, namely tunability τ, is defined as above.

A typical electric field-dependent permittivity is shown in Fig. 2.3. It is measured during the tuning of a $Ba_{0.6}Sr_{0.4}TiO_3$ thin-film varactor. The BST thin-film is between two platinum electrodes at the top and bottom sides. The detailed information regarding the capacitor designs is given in chapter 3.

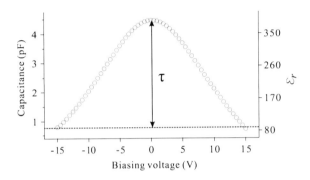

Fig. 2.3 Exemplary tuning of a parallel-plate BST varactor. Measured at room temperature and 1 GHz.

Another critical consideration for microwave applications is the loss in ferroelectric materials. For components built upon such materials, the loss comes from both intrinsic and component related sources. It is the energy exchange between the external electromagnetic wave and the lattice wave of the material, which introduces fundamentally the intrinsic loss [40, 94, 96, 103]. In the meantime, the field induced quasi-Debye loss mechanism also contribute to the loss in microwave frequency range [40, 100], which is dependent on the external field. As an example, in $Ba_{0.6}Sr_{0.4}TiO_3$ the loss depends monotonically on the external field amplitude.

2.4 Piezoelectricity and Electrostriction

Ferroelectric materials in the polarization phase as well as some in paraelectric phase exhibit piezoelectric and converse piezoelectric effects. The piezoelectric effect is the linear interaction between the mechanical and the electrical states in the crystals without inversion symmetry [36]. The piezoelectric effect is a reversible process. On one hand, the crystals accumulate internal charge in response to externally applied mechanical stress, which is the piezoelectric effect. On the other hand, they generate internal mechanical force in response to an external electrical field, which is called converse piezoelectric effect. Piezoelectricity enables applications as sound generation and detection, high voltage generation and frequency generation in either DC or low frequency range. In microwave frequency, especially the converse piezoelectric effect is of interest.

The strain induced in a piezoelectric crystal is defined as the ratio of the deformation to the original dimension. Under certain electric field strength, it can be modeled as in [37]:

$$u_P = \frac{Z(E) - Z(0)}{Z(0)} = d \cdot E , \qquad (2.21)$$

where d is the piezoelectric coefficient, Z is the thickness of a piezoelectric film, E is the external field strength, and u_P is the induced strain.

At the opposite, electrostriction is a property of all dielectric materials, not only ferroelectric materials. It is caused by the presence of randomly-aligned electric domains in the materials. The opposite faces of the domains get charged oppositely when an external electric field is applied across the material. Therefore, the domains attract each other, which results in a reduction of material thickness along the field direction. In [37], the resulting strain u_E is proportional to the square of the field strength:

$$u_E = \frac{Z(E) - Z(0)}{Z(0)} = g \cdot E^2 . \qquad (2.22)$$

where g is the electrostrictive coefficient. Meanwhile, if considering the Poisson's ratio, when a material is stretched or compressed in the field direction, it also tends to compress or expand in the perpendicular directions. Therefore, both longitudinal and transverse deformations shall be observed. Additionally, reversal of the electric field does not reverse the deformation.

It is the electrostriction and converse piezoelectric effect that are typically utilized in microwave transducers. The applications are acoustic wave resonators, e.g. film bulk acoustic wave resonator (FBAR) and surface acoustic wave resonator (SAW). Ferroelectric materials have shown comparatively high electrostrictive effect. More importantly, in ferroelectric materials the permittivity, piezoelectric and electrostrictive coefficients couple with the temperature, external electric field and mechanical strain. These couplings can be modeled as:

$$\begin{aligned} u_{RF} &= s \cdot T_s + [d + g(E_{DV} + E_{RF})] \cdot E_{RF} \\ D &= \varepsilon_0 \varepsilon_r (E_{DC} + E_{RF}) + d \cdot T_s , \end{aligned} \qquad (2.23)$$

where u_{RF} is the strain in microwave frequency range, T_s is the stress, D is the electric displacement, E is the electric field strength, s is the elastic compliance, g is the electrostrictive coefficient, and d is the piezoelectric coefficient [37]. It is obvious that, through electrostriction effect the strain at microwave frequency is partially determined by the external electric field. Actually, for the ferroelectric materials in paraelectric phase, the electrostriction is dominant over the residual piezoelectricity. In other words, the acoustic resonance stimulated by the electrostriction can be switched on and off. A detailed modeling of these electromechanical properties in components is addressed in section 4.1.

2.5 Electrostatic Properties

Besides the dielectric characteristics considered above, the electrostatic properties play an important role in the power consumption and life time of the component based on ferroelectric materials. Though the dielectrics are desired to be ideally isolative, ferroelectrics especially those in thin-film technologies exhibit electric conduction, while in thick-film or bulk ceramics the ceramic thickness and wider electrodes' distance reduce considerably the total leakage current through the material. The ionic transport depends on both the volume and the surface conditions. The charge carriers are first injected from the electrodes through the surface in the volume, and afterwards transport through the volume. In details, the charge injection at the surface is essentially determined by the barrier height on the surface. Additionally the influence of carrier mobility, density of states and permittivity may also contribute to the injection.

The various mechanisms of electron transport have been studied in [55, 68, 99]. The models include thermionic emission, Fowler-Nodheim tunneling, direct tunneling, Poole-Frenkel emission and space-charge-limited current. The Schottky-thermionic emission is one of the dominant mechanisms, especially for the carrier injection into the oxide thin-film with perovskite crystals. There are two factors in the mechanism, i.e. the thermionic emission of the charge carrier through the Schottky barrier and the image force lowering effect, as illustrated in Fig. 2.4(a), which lead to a reduction of barrier height by $\Delta\Phi$ to Φ_B.

Furthermore, when the oxide is charged, e.g. through n-doping, the Schottky barrier is further lowered. As illustrated in Fig. 2.4(b), when there is space charge accumulated at the interface between metal and oxide, the potential of the space charge drops quickly along the depth into the volume. As a result, the barrier is lowered further than that with undoped oxide as in Fig. 2.4(a).

The leakage current J resulted from the Schottky-thermionic emission can be modeled in Richardson-Schottky equation as [88]:

$$J = A^* T^2 e^{-\frac{\Phi_B - \Delta\Phi}{kT}}, \qquad (2.24)$$

and

$$A^* = \frac{4\pi q m^* k^2}{h^3}. \qquad (2.25)$$

where A^* is the effective Richardson constant, T is the temperature, k is the Boltzmann constant, m^* is the effective mass of electrons or holes, h is the Planck constant, q is the elementary charge, Φ_B is the barrier height and $\Delta\Phi$ is the reduction

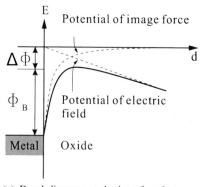 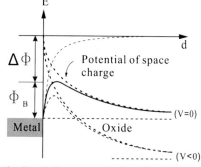

(a) Band diagram at the interface between metal and oxide. The thermionic emission and the image force lowering effect counter each other. Accordingly, the barrier height is lowered by $\Delta\Phi$ to Φ_B.

(b) Band diagram at the interface between metal and n-doped oxide. The potential of space charge further lowers the Schottky barrier.

Fig. 2.4 Band diagram at the interface between metal and oxide

of barrier height as illustrated above. It is obvious that the thermionic emission depends exponentially on the Φ_B and $1/T$. It means that when the barrier is lowered by the image force and the space charge effects, an increase of leakage current is expected.

The barrier height reduction induced by the space charge, or more essentially by the external electric field, can be modeled as:

$$\Delta\Phi = \sqrt{\frac{q^3 E}{4\pi\varepsilon_r\varepsilon_0}}. \qquad (2.26)$$

In order to reduce the thermionic emission, highly resistive layers e.g. SiO_2 or Al_2O_3 can be introduced between the metallic electrodes and the oxide film.

For thin dielectric films, the tunneling mechanism comes into effect. Since there is a quantum process, the mechanism takes place even at relatively low electric field strength [60]. Through an oxide film with thickness lower than e.g. 5 nm the direct tunneling may appear, while on thicker films the Fowler-Nordheim tunneling mechanism turns to be dominant [87]. Therefore, in contrast to the situation with thermionic emission, the very thin buffer film may not efficiently cut the leakage.

Chapter 3
Processing of Ferroelectric Films and Components

Towards the tunable microwave applications, the ferroelectric materials are processed through various techniques and afterwards integrated into components. On one hand, the synthesis and fabrication of materials eventually determine the topology and performance of the components. On the other hand, the design and integration of components shall be adapted and optimized to efficiently utilize the materials' potential. In last decades, bulk and film components have been proposed and investigated. The early research focuses on bulk ceramic process [27, 32]. Afterwards, aiming at higher integration density and lower cost, film components privilege nowadays. The exemplary process can be sol-gel for porous ceramic films, as well as RF magnetron sputtering and pulsed laser deposition (PLD) for thin-film crystals. Typically such processes consist of the paste synthesis, target preparation, deposition, and afterwards combined with components technologies such as photolithography structuring, metalization, passivation and integration etc.

This chapter introduces first the widely used film process technologies for both thick-film and thin-film. The circuit fabrication follows, where the topologies and realization of planar and vertical circuits are addressed. Improvement of component reliability through modification of under-bump materials is then introduced. The characterization and modeling of the lossy components conclude the chapter at the end.

3.1 Processing of BST Thick-Film Components

The term thick-film here refers to the ceramic films with several μm thickness. The manufacture of such films is an additive procedure involving screen-printing process and metalization, where ferroelectric layers and conductors are successively deposited onto an insulating substrate.

3.1.1 Screen-Printed Thick-Film Ceramics

As depicted in Fig. 3.1, the standard process starts with the synthesis of ferroelectric powder of hundreds-nm diameter dielectric crystal grains. A paste is then prepared consisting of the powder, thinners and binders. The paste is then uniformly pushed onto the substrate surface through a woven mesh screen or a stencil by a squeegee. Both the thickness and the pattern of the film are determined in this step. Finally, the paste is dried and sintered in an oven to form connection between crystal grains.

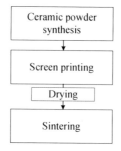

Fig. 3.1 Process of screen-printed thick-film ceramics

Synthesis of BST Powder

An exemplary synthesis of BST ceramic powder is depicted in Fig. 3.2. With standard method, it encompasses the preparation of acetic acid solution and afterwards the sol, freeze drying of the sol, and calcination to make ceramic powder. First the $Ba(CH_3COO)_2$ and $Sr(CH_3COO)_2 \cdot \frac{1}{2}H_2O$ are solved in acetic acid, the ratio between which is determined by the target $Ba_xSr_{1-x}TiO_3$ system. The solution is kept at room temperature under a N_2 atmosphere until the acetates are completely solved. The $Ti((CH_3)_2CHO)_4$ is added then. In order to modify the microwave properties of the final BST film, additionally dopants are added to the

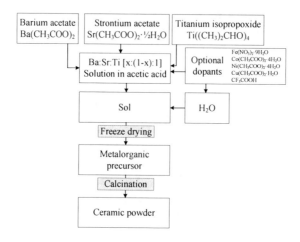

Fig. 3.2 Synthesis of BST ceramic powder

acetic acid solution. The dopants include $Fe(NO_3)_2 \cdot 9H_2O$, $Co(CH_3COO)_2 \cdot 4H_2O$, $Ni(CH_3COO)_2 \cdot 4H_2O$, $Cu(CH_3COO)_2 \cdot H_2O$ and CF_3COOH, in order to introduce the corresponding metallic acceptor and fluorine donor ions. The acceptor and donor concentration can be varied by changing the quotient of dopant solution.

The acetic acid solution is later processed through an exothermal reaction, which yields a viscous sol. The sol is later dried into metalorganic precursor by freeze drying. Water is added to the sol and afterwards the aqueous sol is filtered with a membrane filtered until the PH value is about 3. The aqueous sol is dried by spray drying in a cyclone. Before being collected by the vessel, the dry powder or namely the metalorganic precursor is filter with a 250 µm diameter mesh screen. The metalorganic precursor is calcinated for hours in a tubular furnace under a dry air flow. When the temperature ranges from 900 °C to 1150 °C, the grain size varies accordingly.

Paste Preparation

The calcinated BST powder is then milled within acetone in planetary micro mill, where the 5 mm diameter mill balls are made from magnesia partially stabilized zirconia. In order to separate the acetone and the BST powder then, a rotation carburetor is used. Afterwards, the powder is further fine milled by a attritor mill with water cooling. The mill cup is made of Al_2O_3, while the stirrer and mill balls are made also from Mg stabilized zirconia. The balls are 1 mm diameter. Again the finished powder is separated from acetone by rotation carburetor. The fine powder is then mixed with dispersant by shaft stirrers. The educt is finally homogenized by a three roll mill with Al_2O_3 ceramic mill rolls.

Screen Printing

The screen printing is done by a semiautomatic screen printing machine. The critical process parameters include the distance between the substrate and the screen, the feeding speed of the squeegee, and the pressure of the squeegee. The Al_2O_3 substrate is mounted on printing table by vacuum holder. The prepared paste is squeezed onto the substrate through a woven mesh. The fineness of the mesh determines the smoothness and thickness of the printed film. Fineness refers to the mesh fineness, wire diameter and mesh diameter. The exemplary parameters are $325 mesh/inch$ mesh, 30 µm diameter wire and 50 µm diameter mesh. The printed film is leveled at room temperature under acetone atmosphere, and afterwards in normal atmosphere under cover. The film is dried finally in convection oven at 60 °C. The possible drying cracks of the thick-film can be recovered before sintering by cold isostatic condensation at 300 MPa. As depicted in Fig. 3.3(a), a $Ba_{0.6}Sr_{0.4}TiO_3$ film is printed on a 50 mm×50 mm Al_2O_3 substrate, the printed area is a 48 mm×48 mm rectangular.

In the conceptual heterogeneously integrated circuit, where ferroelectric components may be one part of a system including also

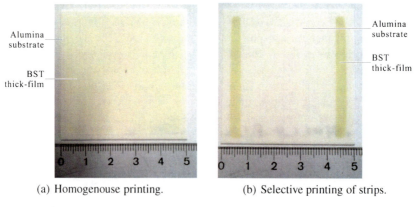

(a) Homogenouse printing. (b) Selective printing of strips.

Fig. 3.3 Screen printed $Ba_{0.6}Sr_{0.4}TiO_3$ on a $50\,mm \times 50\,mm$ Al_2O_3 substrate

microelectromechanical systems (MEMS) or semiconductor components, the patterning of the ferroelectric film is indispensable. Selective printing allows the printing of the thick-film at defined locations. In the standard screen printing process, an additional paste blocking stencil is introduced above the woven mesh. The desired film pattern is then imposed on the substrate instead of a homogeneous coverage, while the remaining steps are maintained as the standard method mentioned above. In Fig. 3.3(b), strips of thick-film are realized on an Al_2O_3 substrate only where the varactors are to be implemented. For the 4 mm wide strip, there is about 0.5 mm wide edges on both sides where the thickness of the film reduces gradually to zero. The slopes on the one hand reduce the resolution of the printing, on the other hand improve the metallic conduction from the circuitries on alumina substrate to the circuitries on top of the film. The selectively printed thick-films are utilized later in section 5.1.

Sintering of Thick-Film

The dry thick-film is composed of ferroelectric ceramic grains and organic residual. In order to clean the residual and make the grains adhere to each other, a sintering step is finally taken. The solid state sintering here is a typical technique to form film from powder by heating the material in a sintering furnace below the powder's melting point until the grains adhere to each other. The method has advantages like preservation of high purity and uniformity of the ceramic precursor, possible control over the binding contact between grains in contrast to melting techniques, good repeatability by control the temperature and atmosphere, and capability of massive production with low cost. A sealed tubular furnace is used here, in connection with a air filtering system. The furnace operates from $650\,°C$ to $1300\,°C$ for various BST thick-films. Since a clean and water free atmosphere is critical for the sintering, especially to keep the concentration of dopant like fluorine, the filtering system includes air dryer, catalytic gas cleaner, and filters with molecular sieves and filling of silica gel, calcium hydroxide and sodium hydroxide, where the steam vapor and

3.1 Processing of BST Thick-Film Components

carbon dioxide produced during sintering are absorbed. The structure of the final film is shown in Fig. 3.4, which is made of $Ba_{0.6}Sr_{0.4}TiO_3$ and sintered for 1 h at 1200°C. The sintering temperature also relates to the composition of the film. In the high temperature cofired ceramics (HTCC) process, the film is sintered at temperature typically beyond 1000°C. But in the low temperature cofired ceramics (LTCC) technique, additives e.g. glass powder in glass matrix composites are introduced in the precursor with the intension to lower the temperature and allow integration of better conductive metal before the sintering. On the one hand, the additives have generally lower melting point, and therefore bind the ferroelectric ceramic grains at a reduced temperature. On the other hand, the additives exhibit lower permittivity and no ferroelectricity which in return reduces the total permittivity and tunability.

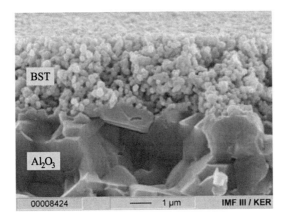

Fig. 3.4 Microscopic structure of the sintered BST thick-film on Al_2O_3 substrate

3.1.2 Planar Circuitry Realization and Reliability

The sinter temperature, i.e. from 650 °C to 1300 °C, constrains the implementation of highly conductive metallic circuitries beneath the thick-film. Therefore planar circuitries are typically realized on top of the thick-film in the form of a single layer metalization. The method is also applicable in the bulk ceramic components, where there is no possibility to insert metal parts in the ceramic volume. The process flow is depicted in Fig. 3.5, illustrating major steps.

The process starts with cleaning of the screen printed thick-film substrate. In order to provide the electric connection over the whole substrate surface for later galvanization, a metallic seed layer is evaporated in vacuum. The layer consists of sequentially 20 nm Cr and 60 nm Au where the thin Cr layer promotes the adhesion between the Au and the porous BST thick-film. Considering the skin depth in the μm range, the reduction of conductivity by the very thin Cr layer is negligible. A positive photoresist is then coated on the seed layer. The seed layer can also be Ni, in order to galvanize Cu later above. Depending on the target thickness of the final metalization,

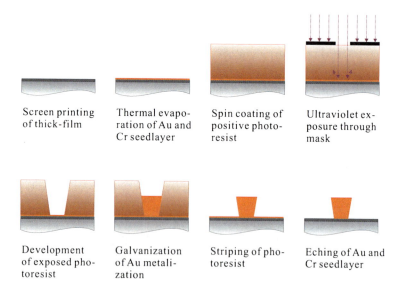

Fig. 3.5 Process flow chart of structuring planar metalization on thick-film ceramics, through photolithography and galvanization

the thickness of the photoresist shall be larger than that of the metal, e.g. 1 µm to 6µm. It is baked under 90 °C to 105 °C before ultraviolet exposure to harden the photoresist. The purpose is to increase the aspect ratio during development of the photoresist. A photo mask is attached to the top of the photoresist, which shades the ultraviolet on the desired locations. The wavelength and power density of the ultraviolet, exposure time, kind of photoresist, roughness of the seed layer, bake temperature and time determine the quality of the exposure. A feature resolution of 3 µm is achievable with the parameters in the Appendix A. The exposed photoresist exhibits about 1000 time faster reaction speed than the unexposed part in alkaline solutions, e.g. the photoresist developer made from KOH solution. Therefore, the high selectivity guarantees the removal of exposed photoresist while reducing the etching of the unexposed part to the minimum. Additionally, oxygen ion etching can also help to clean the atom layer residual in the exposed areas. Afterwards, Au or Cu are galvanized above. The photoresist functions as a mask here. The metalization deposits only in the open areas, while the photoresist coated area is isolated from the electrolyte and therefore unplated. For specific total open area, the galvanization current is to be balanced for both deposition speed and roughness. A high plating voltage introduced by too high current would even damage the photoresist. The photoresist functions as isolation not only in electrolyte but also in acidic etchant by which the seed layer can be removed. Therefore when the target metal thickness is achieved during galvanization, the photoresist mask is removed by either acetone or specific photoresist striper, where the latter kind may provide a smoother edge cleaning. Additional oxygen ion etching can also be applied here for fine cleaning. The plated circuitries with thick Au or Cu metalization are then isolated after the

3.1 Processing of BST Thick-Film Components

removal of the seed layers. An exemplary interdigital capacitor (IDC) realized by this process is shown in Fig. 3.6. The gap width between digits is 8 μm. The Au thickness is 3 μm. The electric field can be established between the neighboring digits where a capacitance is formed. The field penetrates into the ferroelectric film and the substrate. The capacitance through the ferroelectric film dominates the total capacitance. Such IDC is the groundwork of the tunable devices in chapter 5.

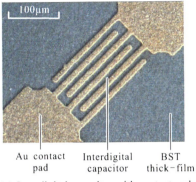
(a) Interdigital capacitor with contact pads.

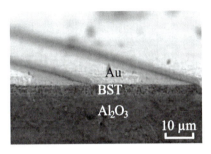
(b) SEM image of the cross section.

Fig. 3.6 Structured interdigital capacitor on BST thick-film

Resistive Bias Network

In the IDC depicted in Fig. 3.6(a), the electric field including both microwave signal and DC bias is established between the metallic digits. Limited by the tunability of $Ba_{0.6}Sr_{0.4}TiO_3$ and the feature resolution in the process, at least 60 V bias voltage is required for a 30 % tunability of the IDC. This bias voltage is still high for most portable applications, e.g. mobile phones, where low cost metal-oxide-semiconductor (MOS) DC-DC converters are limited in their supply voltage. The voltage can be reduced with a narrower gap width while same electric field strength is maintained to achieve the target tunability. However, narrower gap width leads to higher intermodulation distortions when the IDC works at high microwave power level, as well as a higher current density at the edge of the current which increases the metallic loss and reduce the Q factor of the varactor. Therefore, resistive bias network is proposed in order to separate the DC biasing electrodes from microwave signal path. As in Fig. 3.7, high resistive strips made from indium tin oxide (ITO) thin layer are introduced in the gaps between metallic digits [83, 111]. The DC biasing voltage is delivered through the resistive strip marked in dark gray, while the Au electrodes marked in bright gray are DC grounded. The DC electric field is established from the resistive strips to the Au electrodes. As illustrated, the dash-dot lines refers to the DC electric field between resistive strips and Au electrodes and solid lines refer to the RF electric field between Au electrodes.

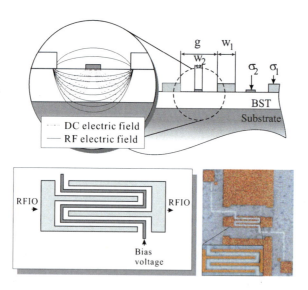

Fig. 3.7 Resistive electrodes for DC biasing in the gap between RF electrodes. σ_1 denotes the conductivity of RF electrodes, while σ_2 denotes that of the resistive electrodes.

However, the insertion of the resistive strips affects the Q factor of the varactor. A comparison between the measured Q factor and finite-difference time-domain method simulations for different strip width w_2 in dependency of the strips' conductivity δ_2 is shown in Fig. 3.8. Due to the low frequency and thin RF electrodes, additional metallic losses are considered in the simulation. It shows a good agreement between simulation and measurement. With strips of high conductivity, the reduction of Q factor comes from the current induced on the strips, which are in serial connection with the capacitance and introduce additional loss. With wider strips, the influence increases. Therefore, only narrow resistive strips are acceptable. At the low conductivity end, less microwave current is induced in the strip. Meanwhile, the metallic loss drops with increased conductivity. These two factors determine that, with either low or high conductivity, the loss is lower than that with the mediate conductivity. The simulation shows a diminishing influence of the resistive strip with very narrow width w_2 while the absolute minimum shifts to lower conductivity.

Prototypes are realized for evaluation. The process is depicted Fig. 3.9. The flow is similar to the one in section 3.1.2. However, in order to introduce the ITO thin layer, additional steps are taken. First, slots are etched in the seed layer where the ITO is deposited directly on the BST thick-film through electron beam or RF sputtering methods. The ITO is then patterned through standard liftoff technique. Afterwards thick Au electrodes are galvanized above. The ITO utilized here is expected to have an effective conductivity of $170\,\mathrm{S}\,\mathrm{m}^{-1}$, which is extracted from a DC resistivity measurement. With a substrate roughness of 300 nm, the measured Q factor of IDC is lower than expected since the radio frequency (RF) conductors are equivalently thinner.

3.1 Processing of BST Thick-Film Components

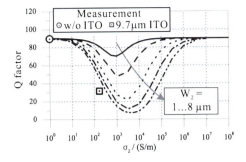

Fig. 3.8 Influence of resistive electrodes' conductivity on the varactor's Q factor

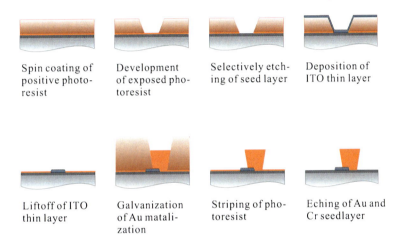

Fig. 3.9 Process flow chart of introducing indium tin oxide in the process in Fig. 3.5

Vertical Connection by Metalized Via

The planar circuitries in section 3.1.2 are easy to fabricate. However, the lack of vertical connections reduces the integration density. As an example, when realizing spiral inductor an internal grounding is indispensable. It can be realized through a long bonding wire which introduces additional parasitics and reduce reliability. Another solution is to form a via by drilling through the substrate and afterward metalizing it, which is flexible in allocation. By controlling the diameter, height and the resistivity of metalization, the parasitics can be controlled. However, considering the Al_2O_3 substrate with BST thick-film printed on one side: if the laser drilling from the BST side, the opening on the front side is larger than the opening on the other side. Considering that the circuitries are typically patterned on the BST side, where small feature size is of higher interest, the laser drills from the bottom side of the Al_2O_3 substrate as illustrated in Fig. 3.10 is preferable. The processing parameters are listed in Appendix A.

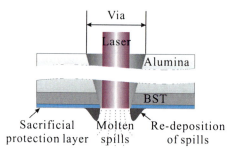

Fig. 3.10 Laser drilling of Al$_2$O$_3$ substrate with BST thick-film. Re-deposition of the molten spills reduces the feature resolution of circuitries.

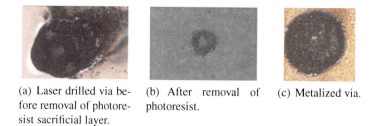

(a) Laser drilled via before removal of photoresist sacrificial layer.

(b) After removal of photoresist.

(c) Metalized via.

Fig. 3.11 Using photoresist layer to protect the thick-film against spills during laser drilling

However, the melting point of BST is about 1350 °C, which is much lower than the 2072 °C melting point of Al$_2$O$_3$ [72]. Therefore, when the laser reaches the BST film, the high temperature molten Al$_2$O$_3$ very quickly melt the BST, and leads to a spill over and redeposition around the via. Therefore, several sacrificial layers have been investigated for the protection of the BST thick-film. As in Fig 3.10, such layer is desired to isolate the redeposited spills. In Fig. 3.11, during the drilling of a 0.3 mm diameter via, the performance of 6 μm photoresist is depicted. The photoresist is first carbonized in a large area around the via as shown in Fig. 3.11(a). It protects efficiently the BST thick-film beneath. After removal of the photoresist as in Fig. 3.11(b), the via shows a clean edge and small foot print. Metallic layers such as Ni, Au and Cr are investigated for protection layer too. However, they melts under the hot spills. The redeposition appears at the openings of the metallic layer and increases drastically the total footprint around the via.

The optimized process is summarized in Fig. 3.12. After the laser drilling, conductive polymer is filled in the via by compressed air from the bottom substrate side. The polymer paste is then curred with hot air. A metalized via is depicted in Fig. 3.11(c). With a 640 μm thick Al$_2$O$_3$ substrate, the via exhibits a lossy inductance. The performance is then modeled as a serial connection of inductor and resistor. With a 0.3 mm diameter, the inductance is 0.35 nH and the resistance is 0.02 Ω. This technique is later utilized in drilling vias in the substrate integrated waveguide filter in section 5.4.

3.1 Processing of BST Thick-Film Components

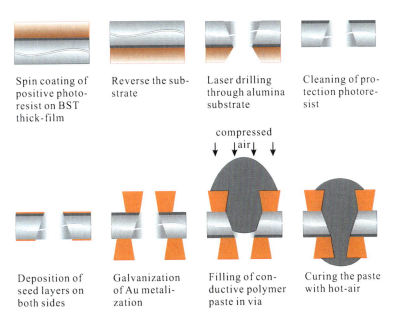

Fig. 3.12 Process flow chart of metalized via through Al_2O_3 substrate and BST thick-film

Component Reliability

The thick-film components can be connected to external circuitries by wire bonding on the carrier circuit board. For higher integration density, they are desired to be integrated into system-in-package (SIP) by flipchip technique, where components are bonded with balls instead of wires. In both cases, the components shall include some contact pad for connection with bonding wires or balls. There is considerable amount of energy loading on the pads during the stud bumping, water cooled dicing and flipchip soldering. Meanwhile, the steps may be conducted under temperature up to 100 °C. Therefore the reliability of the contact pads during the process and operation is of importance. It is the adhesion between the bonding wire or ball and the contact pad, as well as the adhesion between the pad and the BST thick-film, that determines the reliability. A pull test is then conducted under temperature from 70 °C to 100 °C, with adjustable pulling force. As illustrated in Fig. 3.13, a 25.4 μm diameter Au wire is first bonded on the contact pads of a BST varactor. With a hook of the tester, the wire is pulled up with monitored force. There are two modes to measure the adhesion, namely nondestructive mode and destructive mode. The first one monitors the increase of the force till a certain saturation level without destroy the wire or the connection. The latter one keeps on pulling the wire until the force suddenly decreases which is due to either the broken wire or a teared off connection.

Several failure modes are observed as demonstrated in Fig. 3.14. Possible break between the bonding wire and the bonding pad happens at relatively low pull strength. It can be optimized during the bonding process, by e.g. reducing the ultrasonic power or use stud bumping instead of wire bonding. With increased force,

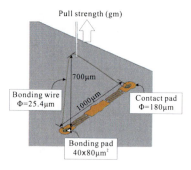

Fig. 3.13 Pull test to evaluate the adhesion of the contact pads

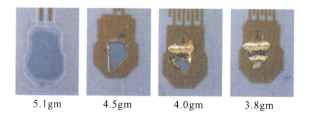

Fig. 3.14 Failure modes of the connection

the connection between the bonding pad of the wire and the contact pad of the varactor gets damaged. It is partially or completely teared off. This has to be coped with more sophisticated methods.

One method is to change the seed layer of the varactor metalization. Typically, Cr and Au are used. Since the Cr can be oxidized gradually by either air or by the BST oxide beneath, the adhesion of the Cr thin layer deteriorates. A more reliable choice of adhesive layer can be Ti. As in Fig. 3.15, the performance of Cr/Au seed layer and Ti/Au seed layer is compared. The Ti/Au seed layer consists of sequential 100 nm Ti layer and 200 nm Au layer. It can be seen that a general improvement of the connection pad adhesion can be achieved.

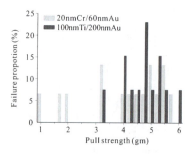

Fig. 3.15 Improvement of connection reliability with Ti/Au seed layer

3.2 Processing of BST Thin-Film Components

The term thin-film here refers to the crystal films below μm thickness. They are typically prepared through vacuum deposition methods, e.g. RF sputtering deposition and PLD. The deposition temperature can be lower than the sintering temperature in the screen printed thick-film in section 3.1.1. Therefore, metallic electrodes can be introduced at the bottom of the ferroelectric layer. Metal-insulator-metal structures, e.g. parallel plate capacitors, are then allowed in these techniques.

3.2.1 BST Thin-Film by RF Sputtering and Pulsed Laser Deposition

The RF sputtering is a sort of physical vapor deposition (PVD) method [70, 93]. In the process, atoms are ejected from a solid target due to bombardment of energetic ionic particles onto the target. The technique is used for thin-film deposition and etching. It provides high deposition speed, low fabrication cost and large processing area. Therefore, it is widely taken in industrial productions. Regarding the quality of deposition, the technique allows targets with high evaporation temperature and good adhesion between the film and the substrate. The produced polycrystalline thin-films have good surface smoothness, flexible thickness and lateral homogeneity. A typical RF sputtering setup is depicted in Fig. 3.16 [34].

The system utilized to deposit the $Ba_xSr_{1-x}TiO_3$ thin-film consists of planar magnetrons, as well as the anode and cathode. The solid target is mounted on the cathode. The cathode is surrounded by a shield, which functions as the anode. On the other side, the substrate is mounted, where the film is to be deposited. During the deposition, the gas e.g. argon is stimulated into plasma by either DC or RF field.

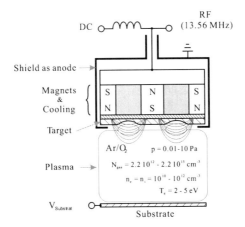

Fig. 3.16 Typical RF sputtering system where the target is located against planar magnetron. Between the target and the substrate, plasma is formed with argon or oxygen gas. Exemplary process parameters are listed as well, where p denotes the pressure, N_{gas} the gas density, n_i the ion density, n_e the electron density, T_e the ion temperature.

The choice of DC or RF field depends on the conductivity of the target material [93]. The ionic particles bombard on the target following the potential drop between the electrodes [70]. The collisions between the incident ions and the atoms in the target exchanges momentum. Collision cascades in the target are then set off by the incident ions. When the collision cascade reach the target's surface with an energy above the surface binding energy, an atom is then shot out [8]. The atom then deposits on the substrate to form the film. Water cooling is indispensable to protect the target from overheating by the bombardment. The permanent magnets mounted above the targets introduce a magnetic field near the target. Through the combination of the magnetic field and electric field in the area, the electrons in the plasma follow a spiral orbit and are constrained close to the target surface. With a long track of motion of the electrons, a higher ionization probability of the gas can be achieved, which reduces the required pressure and voltage. Thereby the atomization rate close to the cathode can be increased by $1 - 2$ orders of magnitude [81]. The disadvantage of the magnetron-anode architecture is that the erosion of the target is not homogeneous, as areas where the field is parallel to the target exhibit higher atomization rate.

The critical deposition parameters include the substrate temperature [2], plasma power [69] and the distance between the substrate and the target [82]. The morphology, structure, crystallization, stoichiometry, electric and dielectric properties can vary under changed deposition conditions [5, 45, 46, 75, 82, 107]. Especially, $Ba_xSr_{1-x}TiO_3$ is sensitive to the deposition parameters. A comprehensive knowledge and precise control of the process are indispensable. A $Ba_{0.6}Sr_{0.4}TiO_3$ thin-film by RF sputtering is shown in Fig. 3.17(a).

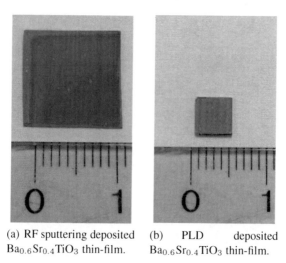

(a) RF sputtering deposited $Ba_{0.6}Sr_{0.4}TiO_3$ thin-film.

(b) PLD deposited $Ba_{0.6}Sr_{0.4}TiO_3$ thin-film.

Fig. 3.17 $Ba_{0.6}Sr_{0.4}TiO_3$ thin-films deposited by RF sputtering and pulsed laser deposition techniques

3.2 Processing of BST Thin-Film Components

Pulsed laser deposition (PLD) is also a kind of PVD deposition method for thin film. In the technique a high power pulsed laser beam is focused inside a vacuum chamber to strike a solid target. The target material is then vaporized into a plasma plume which deposits later on a substrate to form thin-films. It can operate either in ultra high vacuum or with a background gas such as oxygen. It can deposit crystalline thin-films of superior quality than other techniques. It is commonly used to deposit oxide with complex stoichiometry. An exemplary PLD system with integrated reflection high-energy electron diffraction (RHEED) is illustrated in Fig. 3.18. RHEED is a characterization technique of the surface of crystalline materials. RHEED systems focuses only on the surface layer. On the contrary, the transmission electron microscopy and other electron diffraction measurements characterize the bulk sample. It is especially helpful to monitor the PLD deposition process in-situ.

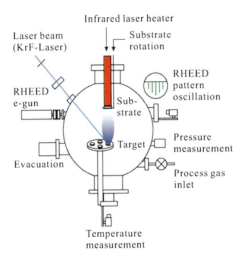

Fig. 3.18 System for pulsed laser deposition with integrated reflection high-energy electron diffraction technique

It is a complex process that the target is first ablated by the laser irradiation and then plasma is formed. The ejection of the atoms from the solid target is achieved by vaporization of the material on the surface in non-equilibrium state. The pulsed laser penetrates into the target within a certain range of depth, which depends on the laser wavelength and the refraction index of the target. Typically the penetration depth is about 10 nm, within which strong electric field is established by the laser to remove the electrons from the target. The electron excitation happens in less than ns time, which is shorter than the period of the laser pulse. The free electrons oscillate in the electromagnetic field induced by the laser and collide with the atoms of the target. The energy exchange from the electrons to the atoms heats up the target surface, which then starts to vaporize [20]. The particles, including atoms, molecules, electrons, ions, clusters and globules, eject into the plasma perpendicular to the target surface towards the substrate, in the form of a plume. The density of the plume reduces when approaching from its center to the edge. The form of plume depends on the pressure in the chamber. With increased pressure, the particles with high energy will be slowed down. The particles with high kinetic energy bombard the substrate, and may re-sputter the deposited film, reducing the deposition rate and resulting in a change of stoichiometry, which hinder the film formation. Therefore, the pressure is one of the critical parameter in PLD process. A $Ba_{0.6}Sr_{0.4}TiO_3$ thin-film by PLD deposition is shown in Fig. 3.17(b). In order to achieve the crystallization, an infrared laser illuminates the backside of the substrate. It heats the substrate to about 900 °C during the BST deposition.

3.2.2 Realization of Thin-Film Components

As depicted in Fig. 3.19, in contrast to the fabrication of thick-film components in section 3.1.2, the fabrication of BST thin-film components starts before deposition of the film. The substrate can be silicon with SiO_2 and TiO_2 overlays, which increases the adhesion between the deposited films and the substrate. As shown in Fig. 3.19, a photoresist mask is structured on top, through the photolithography as that for thick-film components. A typically 150 nm thin Pt layer is deposited above through RF sputtering at room temperature. By removing the photoresist, the desired Pt structures are isolated. BST thin-film is then deposited by either RF sputtering or PLD methods. The typical thickness is from 50 nm to 400 nm. In order to pattern the BST film, one can employ wet etching by a mixture of nitric acid and hydrofluoric acid [113]. Or as depicted, an ion plasma etching is taken. The oxidized photoresist is completely cleaned by oxygen plasma later. Top Pt layer is afterwards deposited by RF sputtering, and patterned by liftoff. Optional Au layer can be finally implemented on the topmost through additional photolithography and galvanization steps. The detailed processing parameters are listed in Appendix A.

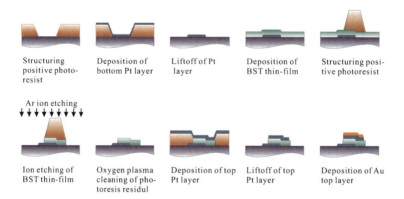

Fig. 3.19 Process flow chart of the BST thin-film components with structured top and bottom electrodes

3.2 Processing of BST Thin-Film Components

The final multilayer structure is analyzed by high resolution electron microscopy (HREM). The stratification is shown in Fig. 3.20(a) where the top Pt layer is not yet deposited. The thickness of TiO_2 is 20 nm, and 300 nm for the SiO_2 layer. The Pt is 150 nm thick, while the $Ba_{0.6}Sr_{0.4}TiO_3$ is 320 nm thick. A good lateral homogeneity is achieved. In Fig. 3.20(b), the root-mean-square surface roughness is measured to below 4.5 nm by atomic force microscope (AFM). The film is then used to fabricate parallel plate capacitors. As shown in Fig. 3.21(a), the capacitor consists of a top Pt electrode and a bottom Pt electrode, in between is the $Ba_{0.6}Sr_{0.4}TiO_3$ thin-film. In the key-hole like structure, it is the rectangular area overlapping the two electrodes that functions as a varactor. A slight misalignment between the top and bottom electrodes does not change the size of the overlap area. Hence, it helps to reduce the requirement of alignment precision.

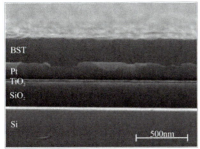
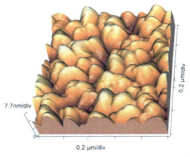

(a) The stratification is taken by a HREM.   (b) Surface smoothness measured by AFM.

Fig. 3.20 The multilayer structure of a BST thin-film component. A Pt bottom layer and a $Ba_{0.6}Sr_{0.4}TiO_3$ thin-film are deposited sequentially on Si substrate.

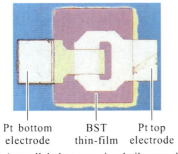
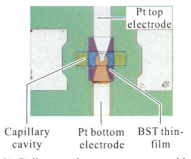

(a) A parallel plate capacitor built upon the BST film.   (b) Bulk acoustic wave resonator with a capillary cavity beneath.

Fig. 3.21 Multilayer components fabricated on BST thin-film

Capillary for Bulk Acoustic Resonator

Ferroelectric materials, e.g. BST, have been proposed for application in FBAR [106]. In such components, the acoustic resonance is established between the top and bottom electrodes, through the piezoelectricity and the induced electrostriction as introduced in section 2.4. The step of the acoustic impedance between the electrode and the BST thin-film results in a reflection of the acoustic wave back into the resonator. Meanwhile a portion of the acoustic energy still leaks into the substrate. Several measures have been proposed to conserve the energy within the thin-film and electrodes. In [76], deep etching from the bottom side of the substrate forms a membrane beneath the bottom electrode. This method however requires both a complex process as well as a direct access to a large footprint on the substrate bottom side. In [106], multiple layers with periodic steps of acoustic impedance have been introduced beneath the bottom electrode. The stratified structure functions as a Bragg mirror, which reflects the energy back. However, as an acoustic resonator, the mirror operates only at the fundamental resonance frequency and harmonics. Therefore a shallow capillary cavity is introduced with the help of a sacrificial layer and a supporting beam as depicted in Fig. 3.22.

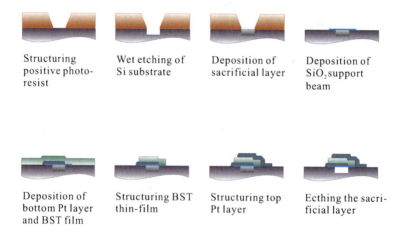

Fig. 3.22 Process flow chart of a ferroelectric thin film acoustic resonator with a capillary cavity beneath

The process is similar to the metal-isolator-metal component process in Fig. 3.19. Additional steps are introduced before the deposition of bottom Pt electrode. The Si substrate with SiO_2 and TiO_2 overlays is first etched from the top with ion plasma etching. A cavity of 100 nm is formed, which is later filled with metallic sacrificial layer, e.g. Cr or Ni. In contrast to the organic sacrificial layers, this metallic sacrificial layer can sustain the high deposition temperature of BST thin-film. However, a direct contact of such metal to the Pt bottom electrode during the BST deposition will lead to alloying, which reduces the conductivity of Pt. Therefore, a supporting beam made from SiO$_2$ is deposited to isolate the sacrificial layer and Pt bottom electrode which is deposited sequentially. Afterwards, BST thin-film and top electrodes are deposited and structured. Finally, when the electrodes are protected with photoresist, the sacrificial layer is etched away by nitric acid. The produced component is shown in Fig. 3.21(b).

3.3 Characterization and Modeling of Lossy Varactors

The varactors fabricated on BST thick-film and thin-film are evaluated for the microwave properties. This characterization is conducted on standard structures, namely characterization kits. The characterization kits are designed generally with simple topologies, which are more tolerant to fabrication uncertainties and numerical extraction procedure of properties. Meanwhile, the same kits are implemented on various films to make the performance comparison easier. The scattering parameters are measured at the tips of wide band ground-signal-ground (GSG) wafer probes by vector network analyzer (VNA). Additional biasing voltage is delivered to the kits through bias-tees which decouple the microwave signal and DC voltage. A typical system covers the frequency range from 40 MHz up to 40 GHz, which is limited by the probes, cables, connectors and dynamic range of the system. In Fig. 3.23(b), IDC varactor is structured on the BST thick-film. Two GSG probes contact the IDC and the ground patches on both ports. The transmission coefficient T is used to extract the impedance of the varactor:

$$T = \frac{2Z_0}{Z_v + 2Z_0}$$
$$Z_v = \frac{2Z_0}{T} - 2Z_0 , \qquad (3.1)$$

where Z_v denotes the varactor impedance and Z_0 is the port impedance of the measurement setup which is 50 Ω here.

In Fig. 3.23(d) and 3.23(f), typical characterization kits for BST thin-films are depicted. For the RF sputtering and PLD deposited thin-films, metal-insulator-metal (MIM) varactors are fabricated with different central pad diameter d ranging from 6 μm to 48 μm. The central pad is distant from the top ground patch at a certain gapwidth. Therefore, the top ground patch has negligible influence on the capacitance measured on the central pad. The tunable capacitance is formed between the octagon shaped pad on the top of BST film and the bottom ground plane below the film. The top pad can be accessed by the signal contact of a GSG probe. The ground tips of the GSG probe contact the large ground patch around the octagon shaped pad, which forms a large capacitance with the bottom ground plane. The capacitance acts as a short circuit at microwave frequency. Therefore most of the RF voltage drops across the small capacitance below the octagon pad area. The bottom ground plane is hot wired to the top ground patch at an edge of the substrate, which allows applying the bias voltage. When several such varactors are

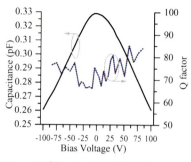
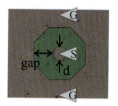

(a) $\tau = 21\%$ under ± 100 V, $Q \leq 65$ with 5 μm thick BST thick-film.

(b) Interdigital capacitor on BST thick-film.

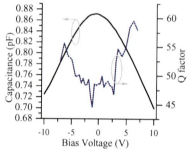

(c) $\tau = 17\%$ under ± 10 V, $Q \leq 45$ with 360 nm thick RF sputtered BST thin-film.

(d) metal-insulator-metal varactor with RF sputtered BST thin-film.

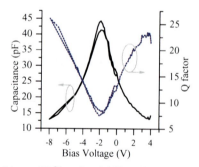

(e) $\tau = 71\%$ under ± 4 V, Q from 7 to 23 with 80 nm thick PLD deposited BST thin-film.

(f) metal-insulator-metal varactor with PLD deposited BST thin-film.

Fig. 3.23 Comparison of the tunability and Q factor of BST varactors from different deposition technologies. All measurements are done at 1 GHz and 25 °C.

3.3 Characterization and Modeling of Lossy Varactors

characterized on same piece of sample, this topology minimizes the influence of the varactor locations. It guarantees a low tolerance of capacitance measured over the whole piece of sample. The reflection coefficient Γ is measured through the GSG probe. The varactor impedance is extracted by:

$$\Gamma = \frac{Z_v - Z_0}{Z_v + Z_0}$$
$$Z_v = \frac{\Gamma + 1}{1 - \Gamma} Z_0 . \quad (3.2)$$

With the complex impedance of the lossy varactor, the capacitance and Q factor are then extracted based on specific models. In order to account the loss in the varactors, resistance is introduced accordingly. As in Fig. 3.24(a), the intrinsic dielectric loss, leakage current and acoustic loss can be modeled as parallel resistance R_p to the ideal reactance $-X_c$, while the metallic loss of electrodes as serial resistance R_s. In the case with a dominant metallic dielectric loss, the model can be simplified to a serial connection of capacitance dielectric and lossy electrodes as in Fig. 3.24(b).

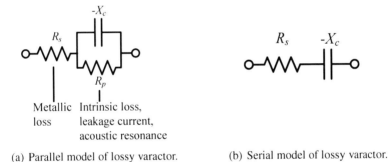

(a) Parallel model of lossy varactor.   (b) Serial model of lossy varactor.

Fig. 3.24 Equivalent circuit models of lossy varactor

In the parallel model, the complex impedance is the combination of all the effects:

$$Z_v = \left(R_s + \frac{R_p X_c^2}{R_p^2 + X_c^2} \right) - j \left(\frac{R_p^2 X_c}{R_p^2 + X_c^2} \right) = R - jX . \quad (3.3)$$

A qualify factor, namely Q factor, can then be defined as the ratio between the stored energy over the dissipated energy. In this case, the Q factor can be determined by:

$$Q \equiv \frac{wX}{wR} = \frac{X}{R} = \frac{\frac{R_p^2 X_c}{R_p^2 + X_c^2}}{R_s + \frac{R_p X_c^2}{R_p^2 + X_c^2}} . \quad (3.4)$$

When the serial resistance is negligible:

$$R_s \ll \frac{X_c^2 R_p}{R_p^2 + X_c^2} \approx \frac{X_c^2}{R_p}, \qquad (3.5)$$

the Q factor is then dominated by the intrinsic properties of the BST thin-film:

$$Q \approx \frac{X_c}{R_p} \approx \frac{1}{\tan \delta}, \qquad (3.6)$$

where the electrodes have no influence on the Q factor any more. If there is no acoustic resonance and the leakage current is negligible, the Q factor is in reverse relation with the dielectric loss tangent defined in section 2.1.

Under the same condition as in Eqn. 3.5, when the metallic loss of electrodes can be neglected, the capacitance can be extracted:

$$C = \frac{1}{\omega \dfrac{R_p^2 X_c}{R_p^2 + X_c^2}} \approx \frac{1}{\omega X_c}, \qquad (3.7)$$

where ω is the angular frequency in the measurement.

By employing Eqn. 3.4 and 3.7, the capacitance and Q factor of the varactors are calculated from the impedance measurements. As compared in Fig. 3.23, there are three phenomena to be addressed.

First, the varactors from different technologies exhibit similar characteristics, including symmetric tuning of capacitance, similar tunability under certain electric field strength. However, due to the different topology, the thick-film varactor requires much higher voltage to establish the sufficiently high electric field strength, e.g. 100 V across 8 μm gap between the digits. In the meantime, the thin-film varactors yield the same electric field strength with only 4 V to 10 V. The voltage reduction is one of advantages of the thin-film varactors.

Second, thick-film varactors exhibit generally higher Q factors than thin-film varactors, even though the stoichiometry of all the films are very close to $Ba_{0.6}Sr_{0.4}TiO_3$, which is measured by X-ray photoelectron spectroscopy. The considerable reduction in Q factor is not induced by the slightly varied stoichiometry, but mainly related to the metallic loss of the electrodes. In Fig. 3.25, an analysis of the parallel circuit model given in Fig. 3.24(a) is illustrated. When assuming a constant 150 nm thickness of the top and bottom electrodes, by varying the metalization from Pt, Al, Au to Cu, the expected Q factor is calculated. It is obvious that the influence of electrodes is considerable. The implementation of better conductors will be helpful, but in the meantime, the compatibility with the processing temperature and oxidizing environment is to be considered. Furthermore, the possible barrier reduction at the interface between electrode and ferroelectric film is also critical concerning component life and leakage current.

Third, in Eqn. 3.4 the relation between the reactance X_c and the R_s also influences the Q factor. With a larger capacitance or smaller reactance, the storage

3.3 Characterization and Modeling of Lossy Varactors 35

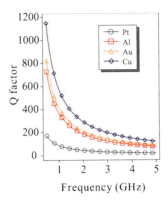

Fig. 3.25 Influence of electrodes' metallic loss on the component Q factor

energy is lower in comparison to the slightly increased dissipation, which leads to a reduced Q factor. This can partially explain why the varactors with thinner film shows lower Q. With the same pad size but a thinner film, the capacitance increases. When it combines with the lossy metalization, the Q factor is lower. In order to identify the effect, several thin-film varactors with different central diameters are realized and measured. The comparison is shown in Fig. 3.26, where all the varacters are fabricated on the same piece of RF sputtered thin-film. With a larger central pad i.e. larger capacitance, the Q factor reduces. By assuming a constant permittivity and loss tangent, which means the capacitance per unit area and the ratio between X_c and R_p kept constant, simulations with the above circuit model confirm that the trend comes from the varied capacitance. In this case, varactors with large capacitance have lower applicable frequency range.

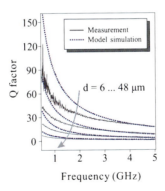

Fig. 3.26 The influence of the electrode pad diameter of BST thin-film varactor on its Q factor

Chapter 4
Novel Multilayer Components on BST Thin-Films

The innovation of the reconfigurable components depends on not only materials but also the circuitry concepts. The reconfigurability of frontends roots in the tunable devices which ground on the fundamental components like varactors and resonators. In this chapter, the multilayer thin-film technology is extended to accommodate novel layers. When a 20 nm thin conductive layer is introduced in the middle of two BST thin-film layers, and antipolarized electric fields are applied to the two layers, the unavoidable generation of acoustic resonance due to the electrostriction can be considerably suppressed. As a result, the degradation of the Q factor due to the acoustic resonance can be reduced. Furthermore, a highly resistive aluminum oxide film is introduced at the interface between BST thin-film and electrodes. When reducing its thickness to 5 nm, an electron injection allows a controllable charge storage on the surfaces of the ferroelectric thin-film, which biases the film without external field. It turns to be a programmable bi-stable capacitor. More importantly, since the thin-film is in paraelectric phase, a low loss is achievable in contrast to the capacitors in ferroelectric phase. Therefore, an innovative bi-stable high frequency capacitor is proposed for the first time.

4.1 Acoustic-Free BST Thin-Film Varactor

As in 3.2.2, BST thin-film varactors are realized in stacked metal-insulator-metal structures. The dielectric layer can be tuned by an external electric field between the top and the bottom electrodes, which reduces the permittivity passively and alters the capacitance value directly. For the efficient use of such varactors in the microwave region, a high Q factor is required which is however affected by the dielectric loss, the limited conductivity of metal electrodes and especially in MIM structures by the acoustic resonances. In contrary to the dielectric and ohmic loss, the loss induced by acoustic resonance is excited by electrostriction only when an external electric field is applied [38]. For typical thin-film structures, where the film thickness is from 200 nm to 400 nm and the electrode thicknesses are below 1 μm, several acoustic resonant modes are excited below 10 GHz. This section starts with characterization of the material properties. It extracts the acoustic impedance and velocity from the measurements, by employing 1D transmission line model and 3D finite element method (FEM) simulation tools. Afterwards a novel multilayer varactor with intermediate resistive bias electrode is proposed, analyzed and verified, which can efficiently suppress the acoustic resonance.

4.1.1 Acoustic Properties

For the design of thin-film tunable MIM capacitors, the narrow band acoustic resonance shall be shifted out of the operation frequency bands. The composition of the stacked layers can be analyzed and optimized, only when their acoustic properties are known. Up to now, the modeling of the acoustic resonance in such ferroelectric varactors is uncommon, not only due to the limited number of software tools which enable the computation of varactor's acoustic response in the microwave domain, but also due to the short of precise knowledge of the acoustic and piezoelectric properties. A 1D acoustic transmission line model of multiple stacked layers is presented in [38, 102]. The extraction can be very efficient. However, since it considers only the material properties, electro-mechanical coupling and responses along the normal surface vector, this method is limited in mapping the real 3D structure.

An exemplary BST thin-film varactor with a single ferroelectric layer is depicted in Fig. 3.20. The BST thin-film is deposited by RF sputtering. The substrate is 100 oriented Si with SiO_2 and TiO_2 layers. 111 oriented Pt and polycrystalline Pt form the bottom and the top electrodes, respectively. By varying the distance d between the target and the substrate as illustrated in Fig. 3.16, the BST thin-film properties including the lateral grain size D, the root-mean-square roughness R_{rms}, the ratio between Ba and Sr, the Ti excess and deposition rate R alter accordingly. The relation is listed in Table. 4.1.

Table 4.1 Dependence of the BST thin-film's properties on the substrate to target distance d during RF sputtering deposition. The properties include the lateral grain size D, the root-mean-square roughness R_{rms} measured by atomic force microscopy (AFM), the Ba/Sr ratio, the Ti excess $y = Ti/(Ba + Sr) - 1$ deduced from Rutherford backscattering spectrometry (RBS) measurements and deposition rate R.

d (cm)	D (cm)	R_{rms} (nm)	Ba/Sr	y	R (nm/min)
5	190	1.7	0.57/0.43	0.26	7
6.3	190	1.8	0.56/0.44	0.27	4.5
7.5	213	2.5	0.57/0.43	0.09	2.4
8.8	135	3.8	0.56/0.44	0.07	1.4
10	108	4.5	0.56/0.44	0.04	0.85

4.1 Acoustic-Free BST Thin-Film Varactor

The realized BST varactors are measured at a controlled temperature of 25 °C. To contact the varactor, GSG probes are used as shown Fig. 3.23(d) and 3.23(f) in an on-wafer measurement setup. In Fig. 4.1 the 1/Q plot illustrates the acoustic resonances raised up by applying bias voltage. The Q factor is measured from 40 MHz up to 16 GHz covering the fundamental and several parasitic resonant modes, namely f_1 to f_5, respectively.

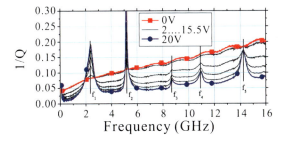

Fig. 4.1 The acoustic resonances raise in the BST thin-film varactor when applying a tuning voltage from 0 V to 20 V. The BST thin-film is prepared at a target to substrate distance of $d = 10$ cm.

The acoustic resonant frequencies and the minimum Q-factor at these resonances have been extracted from the measurements. These characteristics are summarized in Table. 4.2. The tuning voltage is limited to 20 V which results in an electric tuning field strength of 66.7 V/μm within the ferroelectric thin-film.

Table 4.2 Acoustic resonant frequencies and associated Q-factors with a variable target to substrate distance d

d(cm)	f_1(GHz)/Q_1	f_2(GHz)/Q_2	f_3(GHz)/Q_3	f_4(GHz)/Q_4	f_5(GHz)/Q_5
5	2.48/8.4	5.32/3.0	9.23/20.0	11.5/13.1	14.68/6.5
6.3	2.42/8.5	5.37/2.7	9.07/17.8	11.3/11.8	14.60/8.3
7.5	2.40/4.1	5.31/1.4	8.93/14.4	11.30/7.9	14.53/5.3
8.8	2.38/6.2	5.35/2.4	8.79/15.5	11.24/9.2	14.40/7.1
10	2.37/5.2	5.15/1.9	8.70/13.1	10.96/9.9	14.20/5.8

The variations of the resonant frequencies as well as the minimum Q factor at the resonance show a dependence on the process parameter. It can be assumed that the changing morphology of the film results in a change of the films stiffness. The change of the minimum Q factor is related to the change of surface roughness of the films. The observed resonant frequencies f_n do not match with the simple theory of an unloaded acoustic resonator with a length of l_p which corresponds to the BST film thickness and its acoustic velocity c_p:

$$f_n \neq \frac{n \cdot c_p}{l_p}. \tag{4.1}$$

Therefore, a transmission line model of the acoustic resonator is employed to take the influence of the metallic and dielectric layers adjacent to the BST layer into account. It is based on the acoustic transmission line model presented in [38, 102]. The stacked layer structure of the realized BST thin-film varactors is shown in Fig. 4.2. The representation neglects the change of material properties at the interface between two layers while assuming homogeneous properties in each layer. In Table. 4.3, the acoustic properties including the layer thickness of these layers are summarized, except the unknowns of BST film.

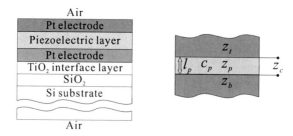

Fig. 4.2 Stacked layer structure of the BST thin-film varactor, and its simplified acoustic transmission line model with various acoustic impedances of the BST thin-film and the adjacent layers

Table 4.3 Acoustic properties of the dielectric and metal layers in BST thin-film varactor [38, 65]. l_{layer} is the layer thickness. c_{layer} is the acoustic velocity. Z_{layer} is the acoustic impedance. Q_{layer} is the acoustic quality factor.

Layer	l_{layer} (nm)	c_{layer} (m/s)	Z_{layer} (kg/m²s)	Q_{layer}
Air	∞	360	400	-
top Pt	400	4230	57.6×10^6	150
BST	l_n	c_p	Z_n	Q_n
Bottom Pt	150	4230	57.6×10^6	150
TiO₂	20	5200	22×10^6	500
SiO₂	300	5100	12.55×10^6	500
Si	300×10^3	8433	19.7×10^6	100
Air	∞	360	400	-

The acoustic impedance at each boundary can be calculated from one of the adjacent layers. For a l long transmission line with a boundary at one end, the input impedance at the other end is calculated as:

$$Z_T = \frac{Z_b + Z_l \tan(\omega l/c_l)}{Z_l + jZ_b \tan(\omega l/c_l)}, \quad (4.2)$$

where Z_b is the boundary impedance, Z_l is the characteristic impedance of the transmission line, Z_T is the input impedance and c_l is the velocity in the transmission line.

4.1 Acoustic-Free BST Thin-Film Varactor

The iterative calculation of the impedance at the lower boundary of the BST film starts at the air interface, by setting Z_b to that of air. The impedance is transformed by the acoustic line formed by the Si substrate into Z_T at the interface between Si and SiO$_2$. This impedance is transformed in a further step by the next layer to the impedance at the next interface, until the interface to the BST film. The same procedure is used to calculate the impedance of the layers on top of the BST film.

The simplified 1D acoustic transmission line is then used to calculate the total input impedance. It consists of the various acoustic impedances and velocities of the BST thin-film and the adjacent layers. In Eqn. 4.3, the impedance Z_n of the varactors is calculated based on the normalized acoustic load impedance z_t and z_b of the top and bottom of the piezoelectric layer respectively as defined in Fig. 4.2 [38]. Due to the simple close form, this model is suited to extract the unknowns of the BST film i.e. acoustic impedance and velocity from wide band measurements with large number of sampling points.

$$Z = \frac{1}{j\omega C}\left[1 - K^2 \frac{\tan\phi}{\phi} \cdot \frac{(z_t + z_b)\cos^2\phi + j\sin 2\phi}{(z_t + z_b)\cos 2\phi + j(z_t z_b + 1)\sin 2\phi}\right] \quad (4.3)$$

where

$$z_t = \frac{Z_t}{Z_p}, z_b = \frac{Z_b}{Z_p}, \phi = \frac{\omega l_p}{2c_p} \quad (4.4)$$

The coupling between the RF electric field and the piezoelectric response of the thin-film is accounted by the field dependent coupling coefficient K^2. It can be calculated by the effective piezoelectric coefficient $d_{33}^*(E)$, the elastic compliance s_{33} and the tunable permittivity $\varepsilon_{33}(E)$ of the film:

$$K^2(E) = \frac{d_{33}^*(E)}{s_{33}\varepsilon_{33}(E)\varepsilon_0 + d_{33}^*(E)}, \quad (4.5)$$

where only the material properties, coupling and responses along the direction of the applied electric field are considered, which are denoted by the index 33.

To reduce the number of unknowns, a constant coupling coefficient K^2 of 0.12 as that under 20 V bias voltage is used in extraction. A parameter sweep is carried out, which varies all unknowns independently in their predicted ranges. For each parameter combination the impedance of the varactor is calculated over a certain frequency range while the frequencies of the arising acoustic resonances are monitored. By minimizing the error between the calculated resonant frequencies and the measured ones, material properties are extracted. Such least mean square method converges with a little mismatch between the extraction and measurement. The approximation is determined by process technologies different to literatures and the tolerance of the assumed materials properties. In addition, the thickness measurement of the Pt electrodes and the BST film has a tolerance of 2 %. To consider all these variations, the number of unknowns increases from 2 to 15 with additional sweep range around the literature values, which is not easily achievable. The extracted acoustic properties of the above mentioned BST thin-films are summarized in Table. 4.4. There is a dependence of the acoustic velocity on the varied process

Table 4.4 Acoustic properties of BST thin-films with varied deposition distance, extracted with the transmission line model

d (cm)	l_p (nm)	c_p (m/s)	$Z_p (10^6 kg/m^2 s)$
5	298	10345	48.4
6.3	341	11046	35.8
7.5	323	11088	36.7
8.8	299	11631	34.1
10	312	11984	36.2

parameter and the associated change in the thin-film morphology. The much higher variation in BST film thickness can be explained by an imprecise film thickness determination using UV-Ellipsometry.

The extracted acoustic properties from the transmission line model are verified by a three dimensional FEM simulation. The commercial solver of ANSYS Multiphysics has been used for coupled field analysis. The model contains the necessary degrees of freedom, i.e. voltages and displacements. The interaction between the coupled electric and stress fields are represented by the piezoelectric matrix. With the fulfilled condition of linear systems, it is able to solve the mechanical displacement under the applied high frequency voltage. Since only the elastic coefficient matrix and the density are accepted instead of acoustic velocity and impedance, necessary interpretations are carried out. The capacitors are modeled with their finite dimensions, the stack of materials and approximate boundary conditions. Hence, the resonant frequencies, not only the fundamental modes, but also the parasitic modes can be predicted at a low error level by the modal analysis [29]. The longitudinal element size is set to be a quarter of acoustic wavelength at the fundamental frequency, which reduces the complexity and may lead to error at high order harmonics. With the material properties in Table. 4.3 and 4.4, the resonant frequencies are comparable with those in Table. 4.2. Their deviation is up to 6 % in fundamental resonant frequencies, and 11 % in higher order harmonics.

It is known that, in stratified structures, the longitudinal strain will couple into the transverse direction [79]. When the BST film is very thin compared to the diameters of contact pads and to the overall size of the continuous BST layer, an

Table 4.5 Comparison of acoustic resonant frequencies between measurement and finite-element-method (FEM) method, all frequencies are in GHz

d (cm)	f_1/f_{1FEM}	f_2/f_{2FEM}	f_3/f_{3FEM}	f_4/f_{4FEM}	f_5/f_{5FEM}
5	2.48/2.44	5.32/5.64	9.23/9.11	11.5/11.1	14.68/15.39
6.3	2.42/2.24	5.37/5.49	9.07/9.26	11.3/10.9	14.6/15.1
7.5	2.4/2.41	5.31/5.9	8.93/9.95	11.3/11.7	14.53/13.5
8.8	2.38/2.48	5.35/6.17	8.79/n.a.	11.24/12	14.4/13.6
10	2.37/2.51	5.15/6.17	8.7/n.a.	10.96/11.9	14.2/13.5

4.1 Acoustic-Free BST Thin-Film Varactor   43

infinite extent of the BST film can be assumed, which suppress transverse resonance. However, transverse propagation is still visible as shown in Fig. 4.3. It leads to reduced accuracy in the extraction of piezoelectric coupling coefficients, which desires further investigation.

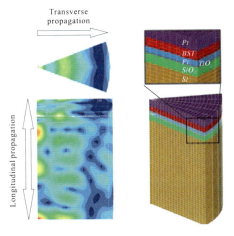

Fig. 4.3 Displacement contour of longitudinal and transverse propagation on stratified structure due to layer stretch. A $10°$ section is simulated with periodic boundary condition, where hotter color denotes larger displacement.

4.1.2 Modeling of Multilayer Acoustic Resonator

To understand and simulate its effect, the modeling of the acoustic resonance is essential for the characteristic of the ferroelectric thin-film. With the following multilayer cancellation methodology, the influence on the capacitance and Q factor can be considerably reduced. First of all, the dual layer resonator can be simplified as in Fig. 4.4. Two ferroelectric layers are deposited sequentially with their individual parameters, including thickness h_L and h_R, as well the electro-mechanical coupling constants e_L and e_R, acoustic impedances Z_{PL} and Z_{PR}, and permittivity ε_{PL} and ε_{PR}, where the subscript L and R denote the parameters for the left and the right films respectively. Here, general electro-mechanical coupling constants are employed, which account for both the piezoelectricity and electrostriction as explained in section 2.4.

Here, the following relation is assumed:

$$\rho_{PL} = \rho_{PR} = \rho_P$$
$$c_{PL} = c_{PR} = c_P , \qquad (4.6)$$

which means the density ρ of the two ferroelectric layers are the same, as well as the stiffness c^E.

In the following, a new dual layer analytic model is proposed. It is extended from the fundamental physical model for one dimensional propagation in [54]. It helps to

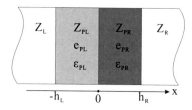

Fig. 4.4 Schematic illustration of a simplified dual layer acoustic resonator with infinitely thick electrodes loading on the ferroelectric layers

identify the physical principles of the stimulation and propagation of acoustic wave in stacked two layer BST thin-films.

In the ferroelectric layers, there is standing wave of the strain, namely u_{PL} and u_{PR}, due the acoustic reflection at the interfaces. In the infinitely thick electrodes, one outgoing propagation is considered for the strain u_L and u_R:

$$\begin{cases} u_{PL} = \beta_{PL-}e^{jk_L x} + \beta_{PL+}e^{-jk_L x} \\ u_{PR} = \beta_{PR-}e^{jk_R x} + \beta_{PR+}e^{-jk_R x} \\ u_L = \beta_L e^{jk_{LL}(x+h_L)} \\ u_R = \beta_R e^{-jk_{RR}(x-h_R)} \end{cases} \quad (4.7)$$

The propagation constant in the left ferroelectric layer is K_L. The amplitudes of the propagating waves are named as β_{PL-}, β_{PL+}, where the propagation in $+x$ direction is marked with $+$ mark, while $-$ for $-x$. In the right ferroelectric layer, the propagation constant is K_R, and the propagating waves' amplitudes are β_{PR+} and β_{PR-}. The outgoing wave constants in the electrodes are K_{LL} for the left one, and K_{RR} for the right one. The amplitudes of propagating waves are β_L and β_R in the left and right electrodes, respectively.

Due to the piezoelectricity and electrostriction, there are electric potentials, namely ϕ_{PL} and ϕ_{PR}, established across the two ferroelectric layers as:

$$\begin{cases} \phi_{PL} = \dfrac{e_{PL}}{\varepsilon_{PL}} u_{PL}(x) + a_{PL}x + b_{PL} \\ \phi_{PR} = \dfrac{e_{PR}}{\varepsilon_{PR}} u_{PR}(x) + a_{PR}x + b_{PR} \end{cases} \quad (4.8)$$

where the constants a_{PL}, b_{PL}, a_{PR}, and b_{PR} are to be determined by the boundary conditions.

Accordingly, the total stresses namely T_{PL}, T_{PR}, T_L and T_R for the left and right ferroelectric layers and electrodes respectively, are determined by the derivative of the strain and the potential:

$$\begin{cases} T_{PL} = j\omega Z_{PL} \left(\beta_{PL-}e^{jk_L x} - \beta_{PL+}e^{-jk_L x} \right) + e_{PL}a_{PL} \\ T_{PR} = j\omega Z_{PR} \left(\beta_{PR-}e^{jk_R x} - \beta_{PR+}e^{-jk_R x} \right) + e_{PR}a_{PR} \\ T_L = j\omega Z_L u_L \\ T_R = -j\omega Z_R u_R \end{cases} \quad (4.9)$$

4.1 Acoustic-Free BST Thin-Film Varactor

Finally the electric displacement D can be determined by:

$$\begin{cases} D_{PL} = -\varepsilon_{PL} a_{PL} \\ D_{PR} = -\varepsilon_{PR} a_{PR} \end{cases} \quad (4.10)$$

Meanwhile, considering the law of electric charge conservation, the following relations hold:

$$\begin{cases} D_{PL} \cdot h_{PL} = D_{PR} \cdot h_{PR} \\ \phi_R \cdot C_R = \phi_L \cdot C_L \end{cases} \quad (4.11)$$

The above eigenmodel of the dual layer resonator is then connected by the boundary conditions:

$$\begin{cases} u_L(-h_L) = u_{PL}(-h_L) \\ u_R(h_R) = u_{PR}(h_R) \\ T_L(-h_L) = T_{PL}(-h_L) \\ T_R(h_R) = T_{PR}(h_R) \\ u_{PL}(0) = u_{PR}(0) \\ T_{PL}(0) = T_{PR}(0) \end{cases} \quad (4.12)$$

In the meantime, the total potential Φ_L and Φ_R across the left and right BST films respectively can be defined as:

$$\begin{cases} \Phi_R = \phi(h_R) - \phi(0) \\ \Phi_L = \phi(0) - \phi(-h_L) \end{cases} \quad (4.13)$$

The above equation system is rewritten as following:

$$\begin{bmatrix} 1 & 1 & -1 & -1 \\ Z_{PL} + [Z_{PL}] & -Z_{PL} + [Z_{PL}] & -Z_{PR} + [Z_{PR}] & Z_{PR} + [Z_{PR}] \\ (z_L - 1)e^{-j\phi_L} & (z_L + 1)e^{j\phi_L} & (z_R + 1)e^{j\phi_R} & (z_R - 1)e^{-j\phi_R} \\ 0 & 0 & \frac{e_{PR}(e^{j\phi_R}-1)}{\varepsilon_{PR}} - \frac{e_{PR}(e^{-j\phi_R}-1)}{\varepsilon_{PR}} + \\ & & j\omega h_R Z_{PR} & j\omega h_R Z_{PR} \\ & & \frac{(z_R+1)e^{j\phi_R}}{e_{PR}} & \frac{(1-z_R)e^{-j\phi_R}}{e_{PR}} \\ \frac{e_{PL}(1-e^{-j\phi_L})}{\varepsilon_{PL}} + & \frac{e_{PL}(1-e^{j\phi_L})}{\varepsilon_{PL}} + & & \\ j\omega h_L Z_{PL} & j\omega h_L Z_{PL} & 0 & 0 \\ \frac{(z_L-1)e^{-j\phi_L}}{e_{PL}} & \frac{(z_R+1)e^{j\phi_L}}{e_{PL}} & & \end{bmatrix}$$

$$\begin{bmatrix} \beta_{PL-} \\ \beta_{PL+} \\ \beta_{PR-} \\ \beta_{PR+} \end{bmatrix} = \begin{bmatrix} 0 \\ 0 \\ \Phi_R \\ \Phi_L \end{bmatrix}.$$

(4.14)

Here the electrode acoustic impedances are normalized to their adjacent ferroelectric layers' impedance as:

$$z_L = \frac{Z_L}{Z_{PL}} \text{ and } z_R = \frac{Z_R}{Z_{PR}}. \quad (4.15)$$

For given materials properties and target frequency, the amplitudes of propagating waves can be solved:

$$\begin{cases} \beta_{PL-} = \frac{[F_r(1-\cos\phi_R)+e^{-j\phi_R}-e^{j\phi_L}\cos\phi_R]\Phi_L^*+[F_l(-1+\cos\phi_R)-e^{j\phi_L}+e^{j\phi_L}\cos\phi_R]\Phi_R^*}{2[-F_r\cos\phi_L+F_l\cos\phi_R+0.5(F_r-F_l)(\cos(\phi_L-\phi_R)+\cos(\phi_L+\phi_R))+j\sin(\phi_R+\phi_L)]} \\ \beta_{PL+} = \frac{[F_l(1-\cos\phi_R)-e^{j\phi_R}+e^{-j\phi_L}\cos\phi_R]\Phi_L^*+[F_l(-1+\cos\phi_R)+e^{-j\phi_L}-e^{-j\phi_L}\cos\phi_R]\Phi_R^*}{2[-F_r\cos\phi_R+F_l\cos\phi_R+0.5(F_l-F_r)(\cos(\phi_R-\phi_L)+\cos(\phi_R+\phi_L))-j\sin(\phi_R+\phi_L)]} \\ \beta_{PR+} = \frac{[F_r(-1-\cos\phi_L)+e^{-j\phi_R}-e^{j\phi_R}\cos\phi_L]\Phi_L^*+[F_l(1-\cos\phi_L)-e^{-j\phi_L}+e^{j\phi_R}\cos\phi_L]\Phi_R^*}{2[-F_r\cos\phi_L+F_r\cos\phi_R+0.5(F_r-F_l)(\cos(\phi_R-\phi_L)+\cos(\phi_L+\phi_R))+j\sin(\phi_L+\phi_R)]} \\ \beta_{PR-} = \frac{[F_r(1-\cos\phi_L)+e^{-j\phi_R}-e^{-j\phi_R}\cos\phi_L]\Phi_L^*+[F_l(-1+\cos\phi_L)-e^{-j\phi_L}+e^{j\phi_R}\cos\phi_L]\Phi_R^*}{2[-F_r\cos\phi_L+F_l\cos\phi_R+0.5(F_r-F_l)(\cos(\phi_R-\phi_L)+\cos(\phi_L+\phi_R))+j\sin(\phi_R+\phi_L)]}, \end{cases}$$
(4.16)

where

$$\begin{cases} F_r = \frac{e_{PR}^2(e^{j\phi_R}-1)}{j\omega\varepsilon_{PR}Z_{PR}h_R} \text{ and } F_l = \frac{e_{PL}^2(1-e^{-j\phi_L})}{j\omega\varepsilon_{PL}Z_{PL}h_L} \\ \Phi_R^* = \frac{\Phi_R e_{PR}}{j\omega Z_{PR}h_R} \text{ and } \Phi_L^* = \frac{\Phi_L e_{PL}}{j\omega Z_{PL}h_L}. \end{cases}$$
(4.17)

Then the strain can be found in Eqn. 4.7.

4.1.3 Suppression of Acoustic Resonances

As in section 2.4, in ferroelectric material the electrostriction exhibits more influence than piezoelectricity when the material operates in paraelectric phase. According to Eqn. 2.23, there are:

$$u_{RF} = s \cdot T + [d + g(E_{DC} + E_{RF})] \cdot E_{RF}$$
$$D = \varepsilon_0\varepsilon_r(E_{DC} + E_{RF}) + d \cdot T_S.$$
(4.18)

The general electro-mechanical coupling coefficient is therefore detailed as:

$$e = d + g \cdot E_{DC},$$
(4.19)

where the DC biasing electric field E_{DC} is assumed much stronger than the microwave electric field strength E_{RF}.

To simplify the problem, it is assumed that the two layers are of same thickness, and anti polarized DC electric field is applied to the layers as:

$$h_L = h_R \equiv h$$
$$E_{L,DC} = -E_{R,DC}.$$
(4.20)

Then by applying Eqn. 4.16 in Eqn. 4.9 at the fundamental resonant frequency ω_0, there is the zero stress on the interface between the ferroelectric layer and electrode. In other words, there is no acoustic vibration propagated to the electrode, and therefore, no energy loss is expected:

$$T_{PL}(-h) = 0$$
$$T_{PR}(h) = 0.$$
(4.21)

4.1 Acoustic-Free BST Thin-Film Varactor

The suppression of the acoustic vibration can be applied to the odd harmonics as well. However, due to the dominant metallic loss at higher frequency as analyzed in section 3.3, the Q factor is low. Therefore, the improvement by reducing the acoustic loss is not essential at harmonics.

Inspired by the above, several demonstrators are made. As depicted in Fig. 4.5, a resistive layer made from 50 nm Pt is introduced between two sequentially deposited $Ba_{0.6}Sr_{0.4}TiO_3$ thin-films to form a resistive strip for DC biasing. A via is opened through ionic plasma etching to connect the top Pt biasing pad and the catch pad of the resistive strip. The bias voltage of the top BST film is $V_T - V_G$, while for the bottom film it is $V_G - V_B$. When setting the V_G as the reference ground and $V_T = V_G$, antipolarized electric field is established in the two layers of BST film. The RF signal is implemented from the top contact to the bottom ground. A prototype is depicted in Fig. 4.6. The capacitance is measured using one port on-wafer GSG probe at the central contact pad. The resistive strip is DC grounded through

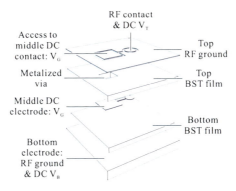

Fig. 4.5 Stratified structure of the dual-layer varactor with resistive middle electrode. The DC voltages are applied from the top DC contact, the middle electrode and the ground separately.

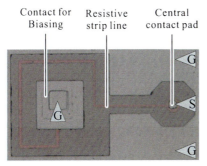

(a) Multilayer varactor with resistive electrode in the middle of two stacked BST thin-films.

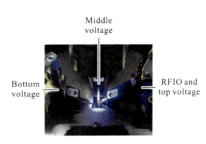

(b) Measurement setup in an automatic probe station.

Fig. 4.6 The prototype with middle resistive electrode and the measurement setup

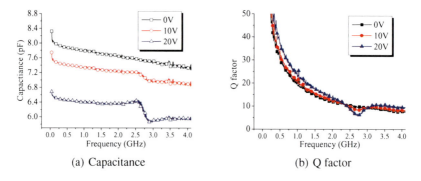

Fig. 4.7 Measured capacitance and Q factor of a BST thin-film varactor. The acoustic resonance appears when applying biasing voltage. The BST thin-film is 310 nm thick.

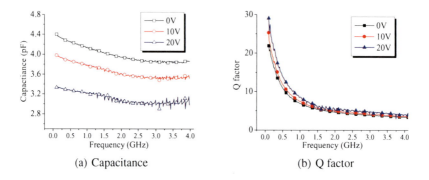

Fig. 4.8 Measured suppression of acoustic resonance in a dual-layer BST thin-film varactor, each thin-film is of 300 nm thickness

the grounding as shown. The pair of anti-polarized DC biasing voltages is delivered through the signal pin of the probe and another pin to the bottom Pt electrode which is not shown in the figure. An automatic on-wafer probe station is employed here for precision measurement which is also shown.

The original capacitance and Q factor of a varactor with a 310 nm single layer BST thin-film is shown in Fig. 4.7. Around 2.8 GHz, the acoustic resonance occurs. The energy exchange from electromagnetic field to mechanical vibrations reduces the Q factor and influences the capacitance as well, which compromise the applicability of such BST thin-film varactors. With proposed dual-layer BST thin-film varactor, the acoustic resonance at the fundamental resonant frequency can be efficiently suppressed, as shown in Fig. 4.8. The fluctuation of capacitance is reduced, while the valley in Q factor disappears. The principle by using multilayer architecture to suppress the acoustic resonance is proven. However, there is a general reduction of Q factor, which is introduced by the resistive middle electrode in two means. First, the middle electrode is in serial connection with the varactors. Therefore the increased equivalent serial resistance lowers Q factor. Secondly, there is no RF choke in the ground pin, which allows leakage of microwave signals to the biasing networks. It further reduces the Q factor. In this case, the improvement by reducing the thickness of the middle electrode and further employing resistive materials such as ITO is to be investigated.

4.2 Programmable Bi-stable Capacitor

There exist in general two ways to alter the capacitance in solid state circuitries. When operating in reverse-biased state, a diode varies the thickness of its depletion zone according to given biasing voltage. Since the capacitance is inversely proportional to the depletion zone thickness, it is controlled by the applied voltage. Diodes show no applicable function of information storage by themselves. In the systems encompassing diodes, all non-volatile memory functions are implemented through external circuitries, which read the specific operation voltages as preset values from additional memory or from real-time calculation. When the external voltage is removed, diodes are reset. In similar means, the capacitance of varactor built on paraelectric materials can be actively varied by external electrostatic field. When operating above their Curie temperature, i. e. in the paraelectric phase of the material, the residue hysteresis is negligible. These varactors rely on external peripherals to store their operation points. The other way is using a ferroelectric material instead of a paraelectric material, where the crystal cell forms semi-permanent electric dipole. By applying external electrostatic field, the dipoles' polarization statistically tends to align with the field, which retains to some extent after removal of the external field. Capacitance hysteresis is therefore introduced as shown in Fig. 4.9. The intrinsic polarization encodes the programmed value without any additional circuitries, and can be detected from the actual capacitance.

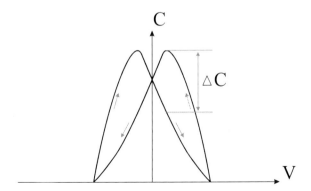

Fig. 4.9 The hysteresis of ferroelectric capacitor can be used for non-volatile memory. The ferroelectric operates below Curie temperature.

However, none of these two ways enables the programmability at microwave frequencies, where low insertion loss is essential for high-frequency capacitors. When operating close to or below Curie temperature, ferroelectric materials show strongly frequency dependent properties in the microwave range. For typical ferroelectric materials such as $Pb(ZrTi)O_3$, the permittivity starts to drop by 50 % between 300 MHz and 3 GHz, while the dielectric loss increases by more than 1.5 times. These lead to high insertion loss in the microwave range. The lack of intrinsic memory function in diodes or in paraelectric material based varactors results in peripheral complexity. Additionally, in diodes a leakage current appears in the reverse-biased state. It increases the power consumption, which turns to be the limiting factor in semi-passive components, e.g. backscattering radio frequency identification (RFID).

This section proposes a passive capacitor with tunable and programmable non-volatile bi-stable capacitance for microwave communications with storage of information. It combines the functions from both paraelectric and ferroelectric materials, and has the advantage that low insertion loss can be achieved in GHz range, while the capacitance varies considerably according to different program status. Meanwhile, the highly resistive buffer layer further reduces the leakage current to 0.01 %, which enables the realization of very low power consuming circuitries. The integration of information storage with tunable capacitance enables an innovative type of passive electronic component, the programmable high-frequency capacitors, which are considered to be applied in the backscatter RFID tags to store the ID numbers, and in the oscillators to program the operation frequency.

4.2.1 Non-volatile Bi-stability by Induced Hysteresis

The capacitor consists of BST thin-film as well as a very thin layer of Al_2O_3 with large energy band-gap and low relative permittivity. The two layers are deposited by RF sputtering sequentially. A schematic illustration is given in Fig. 4.10. When applying external electric field across the Al_2O_3 and BST thin-film, there is a ambipolar charge carrier injection and storage at the interface between the BST thin-film and Al_2O_3 layer. The energy band alignment determined by in situ photoelectron spectroscopy was described in [58]. Because BST and Al_2O_3 have a huge difference in the relative permittivity, a significant part of the total applied voltage drops on the Al_2O_3 layer according to the serial-capacitor structure. Therefore, the energy bands of the Al_2O_3 and the Fermi-level of the Pt electrode shift upwards much faster compared to those of the BST when a negative voltage V_1 is applied. As soon as the Fermi level of the Pt overlaps with the conduction band of the BST, electrons will be injected into the BST by tunneling through the Al_2O_3 layer, since the barrier height at the interface is much too high for the thermionic emission as introduced in section 2.5. Such tunneling process was confirmed by the current-voltage measurements [58] and also observed by other research in similar metal-insulator semiconductor systems [43, 91]. Some injected electrons might be trapped at the defect states and form a negative charge at the interface between Al_2O_3 and BST. This interface charge builds up an internal electric field in both layers and will only then be removed or compensated when a positive voltage (V_2) is applied. At V_2 the

4.2 Programmable Bi-stable Capacitor

flat-band situation is reached in the BST and therefore, the CV curve shows a maximum. If the positive voltage is further increased to V_3, the Fermi level of the Pt will be shifted towards the valence band of the BST and a positive charge will be formed at the interface by means of hole injection. This positive interface charge leads then to a mirrored course of the CV curve and a hysteresis is formed.

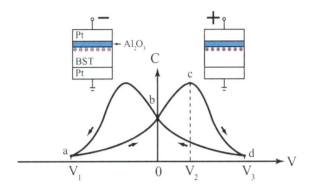

Fig. 4.10 The ambipolar charge carrier injection in the Pt/BST/Al$_2$O$_3$/Pt structure leads to the C-V hysteresis

By assuming the interface charge as a homogeneous charge sheet with an area charge density σ, the boundary condition at both sides of the charge sheet from Gauss law is:

$$\varepsilon_B E_B + \sigma = \varepsilon_A E_A, \tag{4.22}$$

where B denotes that of the BST, and A denotes Al$_2$O$_3$.

When applying a voltage V, the field distribution in the two layer capacitor is given as:

$$V = E_A d_A + E_B d_B, \tag{4.23}$$

where d_A and d_B are the layer thicknesses.

Therefore, the electric field in each layer can be calculated as:

$$E_B = \frac{V - d_A \sigma / \varepsilon_A}{d_B + d_A \varepsilon_B / \varepsilon_A}$$

$$E_A = \frac{\varepsilon_B E_B + \sigma}{\varepsilon_A}, \tag{4.24}$$

where ε is the relative permittivity.

When the internal electric field in BST layer is zero, i.e. $E_B = 0$, the maximal capacitance is achieved. Therefore, the flat-band voltage is given as:

$$V_{FB} = \frac{d_A \sigma}{\varepsilon_A}. \tag{4.25}$$

Since the size of the hysteresis depends on the distance between the two capacitance maxima at $\pm V_{FB}$, therefore it is determined by the thickness of the Al$_2$O$_3$ layer and the density of the interface charge. In addition, the Al$_2$O$_3$ thickness has also an influence on the interface charge σ because the charge injection is based on tunneling process and the degree of the tunneling depends strongly on the electric field. For a certain applied voltage, if the thickness of the Al$_2$O$_3$ increases or equivalently the thickness of the BST decreases, the electric field in the Al$_2$O$_3$ layer increases according to Eqn. 4.23.

Prototype varactors are therefore built. The electrode structure is as same as that in Fig. 3.23(c), and so is the on wafer probe. With varied thickness of Al$_2$O$_3$ layer, different degree of induced hysteresis is observed, according to Fig. 4.11. Stronger hysteresis is measured on the 10 nm thick Al$_2$O$_3$ compared to the varactors with 2.5 nm or the one without Al$_2$O$_3$ layer.

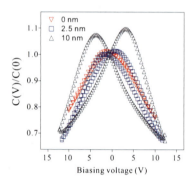

Fig. 4.11 Measurement of induced hysteresis on Ba$_{0.6}$Sr$_{0.4}$TiO$_3$ at 1 MHz, and its dependence on Al$_2$O$_3$ thickness

The induced CV hysteresis based on the bilayer structure offers a bi-stability similar to the ferroelectric materials, but with lower loss because of operating in the paraelectric phase. The switching characteristics of such hysteresis is shown in Fig. 4.12 on exemplary capacitor with a 200 nm Ba$_{0.1}$Sr$_{0.9}$TiO$_3$ which is completely in paraelectric phase at room temperature, with 5 nm Al$_2$O$_3$ layer. The operation contains basically a writing and a reading process, which are carried out with the corresponding write voltage V_{Write} and read voltage V_{Read}. As shown, the capacitor is preprogrammed with a short DC voltage (15 V, duration 2 s). The CV curve takes then the lower branch of the hysteresis at positive voltages and a low capacitance can be read out with the read voltage (2 V) correspondingly. If a write voltage with the opposite polarity (−15 V) is applied, the CV curve will take the upper branch and a high capacitance can be measured. For both cases the capacitance was measured continuously over time and does not vanish after the removal of the write voltage, showing clearly a non-volatile repeatable programmability. The choice of writing and reading voltages is in principle arbitrary. However, writing voltage should be larger than the flat-band voltage V_{FB} in order to generate a high charge carrier

4.2 Programmable Bi-stable Capacitor

injection and therefore a high interface charge. The reading voltage V_{Read} should be between 0 and V_{FB} but preferably close to V_{FB} in order to get a large difference between the high and low states. The polarity of writing and reading voltages can be arbitrarily correlated since the hysteresis is symmetric.

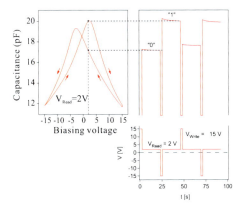

Fig. 4.12 C-V and switching characteristics at 1 MHz of a $Ba_{0.1}Sr_{0.9}TiO_3$ capacitor with 5 nm Al_2O_3. The voltages for the write and read operation are ± 15 V and 2 V, respectively.

The frequency dependence of the memory window is further measured up to 2 GHz, and compared to the window at 1 MHz as shown in Fig. 4.13. The memory window keeps relatively constant across the wide frequency range. The frequency dependence of Q factor is also shown. Considering the dominant metallic loss discussed in section 3.3, the reduction of Q factor at high frequency is not due to the induced hysteresis. Therefore, such bi-stable capacitors show higher Q factors than ferroelectric varactors.

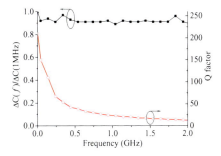

Fig. 4.13 Frequency dependence of the memory window and Q factor of the bi-stable capacitor. The memory window is normalized to the one at 1 MHz. The shown Q factor is the minimal value under tuning voltage from 0 V to 10 V at each frequency points.

4.2.2 Long-Term Stability

The retention property of the BST/Al$_2$O$_3$ capacitor was characterized with an extended time scale, as demonstrated in Fig. 4.14. The stability of the charge storage is crucial for the applications as passive programmable high-frequency devices. The interface charge can disappear gradually through back tunneling or recombination. For such measurements the capacitor is pre-programmed with ± 15 V at the beginning, the corresponding capacitance is measured under 2 V and at logarithmically distributed sampling time points. The initial difference of the capacitance between the high and low states is normalized to 1. At the beginning the two capacitance states approach each other rapidly, but get more stable after approximately 100 s. The low state shows a comparably more stable behavior because its read voltage in this case has the same polarity as the interface charge. The energy band under this condition favors the stabilization of the trapped charge carriers. By extrapolating the lines logarithmically [56, 63, 66, 92] the residual difference capacitance between the two states is estimated to be 15 % in 1 year and 6 % in 10 years. Both values show a promising potential for the long-term stability.

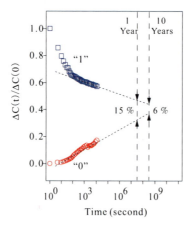

Fig. 4.14 Typical retention characteristics of the bi-stable capacitor with 5 nm Al$_2$O$_3$. The normalized memory window which is the difference in capacitance between the high and low state is extrapolated to 1 and 10 years at room temperature.

Chapter 5
Tunable Multiband Ferroelectric Devices

Future wireless telecommunications systems will have to seamlessly operate across multiple radio technologies. The mobile terminals shall support a large variety of wireless standards, which allow ubiquitous connectivity through horizontal and vertical handovers [50, 62]. The challenge of multiband and multi-standard operation is expected to be met by the frequency-agile, cognitive and software-defined radio (SDR) In this concept, a shared hardware platform shall be utilized for different standards. The inspection of the various standards reveals that a large variety of system parameters has to be supported [31]. As depicted in Fig. 5.1, similar to the reconfigurability at the baseband, where modulation scheme, time division, channel equalization and protocol of processing software are involved, a dynamical reconfiguration of the RF frontend, the system's operation frequency, bandwidth, power level, as well as spatial diversity can be adapted to cope with the time and regional variations of traffic demands, meanwhile a high efficiency can be stabilized under the changing environment [44].

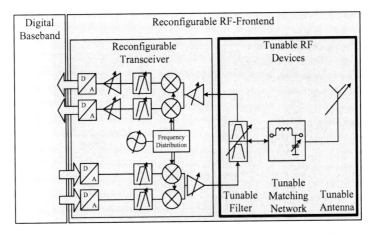

Fig. 5.1 Reconfigurable architecture of multiband software-defined radios

Y. Zheng: *Tunable Multiband Ferroelectric Devices*, LNEE 228, pp. 55–136.
DOI: 10.1007/978-3-642-35780-0_5 © Springer-Verlag Berlin Heidelberg 2013

The progress in semiconductor technology has shown the capability in implementing flexible platforms in system on chip (SoC). But the demands on high performance multipurpose devices at low cost together with the challenge on low energy consumption are driving increasing research efforts in a variety of other enabling technologies, such as MEMS, piezoelectric and ferroelectric components. By utilizing the fundamental components like ferroelectric varactors, which are addressed in the previous chapters, the reconfigurability of frontends can be enabled by novel devices including the tunable multiband antenna, tunable matching network and tunable substrate integrated waveguide filter. They are addressed in this chapter. The challenging trend towards compactness with wider spectrum coverage of an antenna is coped with several tunable resonant modes, which on one hand provide narrower stationary bandwidth than the static counterparts, on the other hand, they can be tuned to the desired frequency and equivalently increase the spectrum coverage to allow multiband operations. The environmental impact and frequency dependency of antennas can be compensated by the tunable matching network. Through an efficient control, it can not only stabilize the antenna impedance, maintain the transducer gain, but also increase the frequency coverage. The complete design methodology for single band matching networks is proposed, and extended to multiband designs. In the meantime, varactor loss is considered in both the circuit and transmission function levels, which helps to efficiently reduce the insertion loss. An evanescent mode substrate integrated wave cavity is loaded with a pair of complementary split ring resonators. The ring resonators are tuned through embedded varactors. The input mismatch due to impedance drift during tuning is compensated by integrated tunable matching networks. The investigation of the devices generally roots in the analytical modeling and fullwave simulation methods, by which their theoretical performance boundaries are identified. The requirements on BST varactor's tunability and loss are afterwards recognized. The designs are verified with demonstrative prototypes, where processing technologies of BST films are adapted.

5.1 Tunable Multiband Antennas

As shown in Fig. 5.2(a) nowadays embedded wireless modules tend to provide multiple services, while pursuing compact dimensions. The trend is challenging the commonly used static antennas. The compactness demands not only antennas but also chassis of small volume. It consequently increases the radiation quality factor as illustrated in Fig. 5.2(b). Due to the fundamental limitation on the attainable frequency bandwidth [35], the further compactness results in a reduction of the bandwidth and therefore, invalidates the implementation of single wide band antennas. In addition, the antennas are required to utilize efficiently the restricted volume and chassis footprint. In the multiband antennas where several dedicated resonant structures are complementary, a trade-off between the conciseness and the simultaneous frequency coverage is inevitable. Tunable antennas have proven to balance the demands, by tuning a narrow instantaneous band to cover a wide frequency range. A cost effective way to tune an antenna is to load it with reactive tunable components e.g. varactors. Varactors of various technologies have been implemented,

5.1 Tunable Multiband Antennas

including diodes, MEMS and ferroelectric varactors [3, 6, 42, 80, 114]. In [115] low-cost ferroelectric varactors based on Barium-Strontium-Titanate (BST) thick-film technology have been reported to efficiently suppress the harmonic radiation. multiple chart:

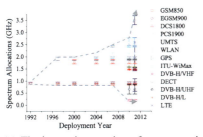 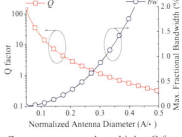

(a) The increasing number of spectrum allocations for multiple services.

(b) Compact antennas have higher Q factor and therefore, narrower stationary bandwidth. A is the antenna's aperture, λ is the operating wavelength.

Fig. 5.2 The demands on both wider spectrum coverage and further compactness impose challenges on the multiband antennas

In this section, a detailed theoretical analysis quantifies the benefit of using tunable antennas instead of static counterparts, as well as the requried minimal tunability of the varactors. A capacitively loaded antenna is then modeled for its nonlinearity. The suppression of harmonic radiation by using ferroelectric varactors is addressed. Demonstrative single-band and multiband tunable antennas are then presented. The concepts are finally extended to a fully integrated ceramic antenna optimized for frequency division duplex (FDD) services.

5.1.1 Theoretical Analysis: Q Factor, Bandwidth and Tunability

A tunable antenna is an extension of static antennas. The tunable reactive loads on one hand lower the center frequency and reduce the stationary bandwidth, while on the other hand expand coverage of the center frequency which can be equivalently considered as a bandwidth enhancement, but in a dynamic way. The following analysis begins with the introduction about the relation between the bandwidth and the Q factor in static antennas. It extends later the theory to take the varactors into account, and comes to the incidence relations between the reactive load, stationary bandwidth, dynamic bandwidth, miniaturization and Q factor.

The key performance indicators of a static electrically small antenna are the stationary bandwidth, total efficiency, and the compactness. Generally, antenna's bandwidth is determined when the frequency-dependent input impedance is known. In detail, the bandwidth can be defined as the reflection coefficient below a certain threshold, e.g. -6 dB for compact antennas. Through the implementation of matching networks, the bandwidth can be enhanced. Without loss of generality, the bandwidth definition from Bode-Fano matching theory is taken [35, 78]. It provides the fundamental constrains on the reflection coefficient when using arbitrary realizable matching networks.

The following analysis extends the relation between the bandwidth and the Q factor of varactor loaded narrow band antennas.

As the center of the analysis, it is the Q factor that bridges the dimensions and bandwidth. In ideal case, Q factor is defined as the ratio between the energy stored in the reactive near field and the radiated power:

$$Q = \frac{2\omega \max(W_m, W_e)}{P}, \quad (5.1)$$

where W_m and W_e refer to the stored magnetic energy and the stored electric energy respectively, and the ω is the angular frequency [21, 24]. At the resonant frequency, namely ω_0, the inductive part and capacitive part are equal.

When the antenna is circumscribed within a spheric surface of radius a, it is in the evanescent field in the near range that the non-radiated energy is stored. Therefore, the Q factor is finally determined by the geometries as in [24]:

$$Q = \frac{1}{k_0 a} + \frac{1}{(k_0 a)^3}. \quad (5.2)$$

When the antenna is miniaturized while keeping the operating frequency, the Q factor increases. However one shall notice that, the reactive energy is considered here in the near field, which desires a revision when lumped elements are introduced.

The Q factor provides a simply way to quantify the bandwidth. In [41], by employing a second order band approximation of the reflection coefficient, it is shown that the Bode-Fano bandwidth of a narrow-band antenna is determined by the amplitude of the frequency scaled frequency derivative of the reflection coefficient, namely $\omega_0 |\Gamma'|$ where ω_0 denotes the center angular frequency and Γ' the frequency derivative of the reflection coefficient. More importantly, the Q factor definition is identified as the Q factor of a first order accurate approximating resonance model of the antenna:

$$Q_\Gamma = \omega_0 |\Gamma'| = \frac{\omega |Z'|}{2R}, \quad (5.3)$$

when the Q factor is sufficiently larger, or in other words the antenna has a narrow bandwidth.

5.1 Tunable Multiband Antennas

The corresponding maximal attainable bandwidth at reflection threshold Γ_0 from Bode-Fano theory is then given by:

$$B = \sqrt{Q^2 K_0^2 + 4} - QK_0 \approx \frac{\pi}{Q \ln \Gamma_0^{-1}} + O\left(Q^{-3}\right),$$

where K_0 stands for $2 \ln \Gamma_0^{-1}/\pi$. For antennas with stationary narrow bandwidth, it can be further simplified as:

$$B \approx \frac{27}{Q |\Gamma_{dB}|}. \tag{5.4}$$

In the following analysis, there are two assumptions. First, the antenna's unloaded stationary bandwidth is narrow enough to hold the approximation in Eqn. 5.4. Second, the antenna is considered as a linear component. The nonlinear behavior is discussed in chapter 5.1.2. Therefore, a first-order lumped equivalent-circuit model in Fig. 5.3 is utilized. It is in a parallel RCL form.

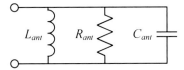

Fig. 5.3 First-order lumped equivalent circuit of a single-band antenna

The antenna's intrinsic inductance L_{ant} and capacitance C_{ant} are related to the unloaded Q factor namely Q_{ant}:

$$C_{ant} = \frac{Q}{R\omega_0} \text{ and } L_{ant} = \frac{R}{Q\omega_0}. \tag{5.5}$$

Based on the above modeling, single-band tunable antenna can be extended as in Fig. 5.4(a). When the varactor loads on the antenna, its position determines the effective load at the antenna's input port. The effect is modeled here using a transformer. Considering the above mention assumption on linear system, the actual varactor C_v can be represented as C_{load}, whose capacitance is determined as:

$$C_{load} = |n|^2 C_v, \tag{5.6}$$

where n is the complex voltage transformation ratio between the RF voltage across the varactor and that at the feed port. The capacitor loads directly at the port in parallel to the intrinsic resonator as in Fig. 5.4(b). In the following only a narrow tuning range of the varactor and the operating frequency is considered, where n is assumed to be constant.

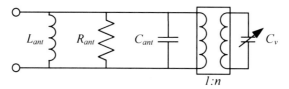

(a) the actual varactor couples to the resonator through an equivalent transformer.

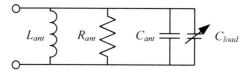

(b) effect capacitive load in parallel to the resonator.

Fig. 5.4 Equivalent circuit of a capacitively loaded tunable antenna

According to Eqn. 5.3, for ideal lossless varactors, the loaded Q factor can be determined as following:

$$Q_T = \frac{V_{in}^2/X_{ant} + V_{in}^2/X_{load}}{V_{in}^2/R_{ant}} = \frac{R_{ant}}{X_{ant}} + \frac{R_{ant}}{X_{load}} \equiv Q_{ant} + Q_{load}, \quad (5.7)$$

where X_{ant} is the minimal intrinsic reactance and X_{load} is the equivalent load's reactance.

Here Q_{load} is not the Q factor of the load but the ratio between the load reactance X_{load} and the radiation resistance R_{ant}, which represents the shift of antenna's Q factor of the antenna due to the load. It is related to the unloaded Q factor as following:

$$\frac{Q_{load}}{Q_{ant}} = \frac{X_{ant}}{X_{load}}. \quad (5.8)$$

The resonant frequency of the antenna is determined by the total reactance. In Eqn. 5.9, the frequency ω_0 without load and the frequency ω_1 with load are considered:

$$\omega_0 = \frac{1}{\sqrt{L_{ant}C_{ant}}}$$
$$\omega_1 = \frac{1}{\sqrt{L_{ant}C_{total}}} < \omega_0. \quad (5.9)$$

In the case of the parallel RCL model, at arbitrary frequency ω, the intrinsic reactance and the load reactance are:

5.1 Tunable Multiband Antennas

$$\frac{1}{X_{ant}} = \frac{1}{\omega L_{ant}} + \omega C_{ant} = \frac{\omega_0^2 + \omega^2}{\omega} C_{ant}$$

$$\frac{1}{X_{load}} = \omega C_{ant}. \qquad (5.10)$$

By applying these into Eqn. 5.8, there is:

$$\frac{Q_{load}}{Q_{ant}} = \frac{\omega^2 C_{load}}{(\omega_0^2 + \omega^2) C_{ant}}. \qquad (5.11)$$

Accordingly, the Q factor shifts, which is represented by Q_{load} at the frequency ω_1:

$$\frac{Q_{load}}{Q_{ant}} = \frac{C_{load}}{\left[(\omega_0/\omega_1)^2 + 1\right] C_{ant}}. \qquad (5.12)$$

Now the loading effect of the varactor can be simply defined by a loading factor Lo as:

$$Lo = \frac{C_{load}}{C_{ant}}. \qquad (5.13)$$

Then, the Q factor shift is directly related to the loading factor:

$$\frac{Q_{load}}{Q_{ant}} = \frac{Lo}{(\omega_0/\omega_1)^2 + 1}. \qquad (5.14)$$

In the following, analysis on the miniaturization, frequency tuning and bandwidth enhancement are to be addressed. Some frequently used notations are listed in Table 5.1, which are further illustrated in Fig. 5.5.

Table 5.1 Notations in the following analysis

f_0	unloaded center frequency
f_1	untuned loaded center frequency
f_1'	tuned loaded center frequency
bw_{s_0}	unloaded fractional stationary bandwidth
bw_{s_1}	untuned loaded fractional stationary bandwidth
bw_{s_1}'	tuned loaded fractional stationary bandwidth
bw_d	fractional dynamic bandwidth
τ_{C_v}	tunability of varactor
τ_f	tunability of center frequency

Miniaturization

The compactness of capacitively loaded antenna is of a focus. With certain dimensions, the compactness can be interpreted as the ratio of the original frequency without load to a lower one with load. When the antenna is circumscribed within a spheric surface of radius a, its electric dimension at the frequencies ω_0 and ω_1 can be related as:

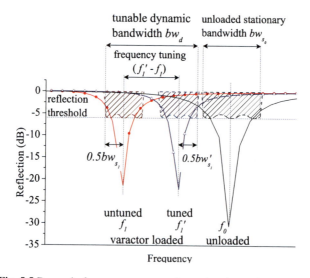

Fig. 5.5 Dynamic frequency coverage by tuning the stationary band

$$\frac{a}{\lambda_1} = \frac{a\omega_1}{2\pi C} < \frac{a\omega_0}{2\pi C} = \frac{a}{\lambda_0}. \tag{5.15}$$

By taking the ratio between them, a miniaturization factor η_m is defined as:

$$\eta_m \equiv \frac{\omega_0}{\omega_1}. \tag{5.16}$$

By applying in Eqn. 5.14, it can be simply related to the Q factor shift as:

$$\frac{Q_{load}}{Q_{ant}} = \frac{Lo}{\eta_m^2 + 1}. \tag{5.17}$$

The frequency shift with load as in Eqn. 5.9 is then related to the loading factor:

$$\left(\frac{\omega_0}{\omega_1}\right)^2 = \frac{C_{total}}{C_{ant}} = 1 + \frac{C_{load}}{C_{ant}} = 1 + Lo, \tag{5.18}$$

i.e. the miniaturization is determined by the loading factor. When varying the η_m, the influence on compactness can be found in Fig. 5.6(a):

$$\eta_m^2 = 1 + Lo. \tag{5.19}$$

Using Eqn. 5.14 and 5.18, the Q factor shift is found to be linearly related to the loading factor as:

$$\frac{Q_{load}}{Q_{ant}} = \frac{Lo}{2 + Lo}. \tag{5.20}$$

5.1 Tunable Multiband Antennas

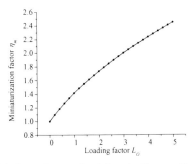
(a) dependence of miniaturization on the loading factor.

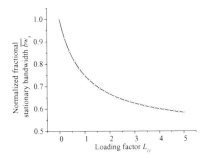
(b) dependence of stationary bandwidth on the loading factor.

Fig. 5.6 Influence of capacitive load

The attainable bandwidth can be calculated from the Eqn. 5.7 for Q_{total} and Eqn. 5.4 as illustrated in Fig. 5.6(b), where $\overline{bw}_s$ is normalized bandwidth bw_{s_1}/bw_{s_0} at a same reflection threshold Γ_{dB}.

It is concluded that, by increasing the capacitive load, the stationary bandwidth reduces while the operating frequency is lowered, or in other words the antenna is miniaturized. The intrinsic components of the first order equivalent circuit are assumed here to be independent on frequency. However, this prerequisite shall be reconsidered when dealing with heavy loads and hence large frequency shift.

Frequency Tuning

When the capacitance of the load alters, it tunes the resonant frequency since the total capacitance in the resonator is tuned. Naming untuned capacitance as C_{load}, tuned capacitance as C'_{load} and assuming $C_{load} > C'_{load}$, the relative change of the total capacitance $\tau_{C_{total}}$ is given as:

$$\tau_{C_{total}} = \frac{C_{load} - C'_{load}}{C_{load} + C_{ant}} = \frac{\Delta C_{load}}{C_{load} + C_{ant}} = \frac{\tau_{C_{load}}}{1 + 1/Lo} . \quad (5.21)$$

Meanwhile, the frequency increases from f_1 to f'_1. The tunability of the frequency can be defined as τ_f:

$$\tau_f = \frac{f'_1 - f_1}{f_1} = \frac{1/\sqrt{L_{ant}C'_{total}} - 1/\sqrt{L_{ant}C_{total}}}{1/\sqrt{L_{ant}C_{total}}} = \sqrt{\frac{C_{total}}{C'_{total}}} - 1 . \quad (5.22)$$

If $\tau_{C_{total}} \ll 1$ which refers to a low load tunability, the frequency tunability can be approximated as:

$$\tau_f = (1 - \tau_{C_{total}})^{-\frac{1}{2}} - 1 \approx \frac{1}{2}\tau_{C_{total}} . \quad (5.23)$$

It relates the frequency tunability and load tunability. Since the tunability of the load is identical to that of the varactor according to Eqn. 5.6 i.e. $\tau_{C_v} = \tau_{C_{load}}$, the frequency tunability is determined finally by:

$$\tau_f = \frac{\tau_{C_{load}}}{2(1+1/Lo)} = \frac{\tau_{C_v}}{2(1+1/Lo)}. \tag{5.24}$$

An indication of how efficiently the antenna utilizes the varactor during tuning is defined, namely the frequency tuning efficiency η_t:

$$\eta_t \equiv \frac{\tau_f}{\tau_{C_v}} = \frac{Lo}{2(1+Lo)}. \tag{5.25}$$

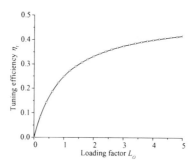

Fig. 5.7 The dependence of frequency tuning efficiency on the loading factor

It is clear that the frequency tunability is always below a half of the varactor's, and by increasing the loading factor, the antenna is prone to be more tunable.

Dynamic Bandwidth

As in Fig. 5.5, by shifting the operation band, the tunable antenna covers a wide frequency range. In typical services like global system for mobile communications (GSM) and universal mobile telecommunications system (UMTS), only narrow channels are allocated to each client for uplink and downlink. In this case, the antenna can cover the channels with narrow but dynamically tunable bands. The overall frequency coverage is then treated as dynamic bandwidth.

The above analysis has shown that by adding capacitive load, the stationary bandwidth is reduced, while the resonant frequency is shifted. As in Fig. 5.8, the first factor reduces the frequency coverage, while the latter one compensates by covering dynamically the frequency range. These two opposing factors determine the overall coverage, or namely the sum of stationary bandwidth and frequency tuning range.

To take the change of stationary bandwidth during tuning into account, the stationary bandwidth at the lower and higher tuning boundary, bw_{s_1} and bw'_{s_1} respectively, is determined by:

5.1 Tunable Multiband Antennas

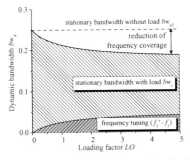
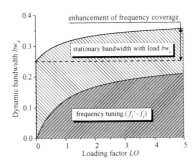

(a) Dynamic bandwidth is reduced with 10 % varactor tunability τ_{C_V}.

(b) Dynamic bandwidth is enhanced with 50 % varactor tunability τ_{C_V}.

Fig. 5.8 Dependency of the fractional dynamic bandwidth on both loading factor Lo and varactor tunability τ_{C_V}

$$bw_{s_1} = \frac{bw_{s_0}(2+Lo)}{2(Lo+1)}$$

$$bw'_{s_1} = \frac{bw_{s_0}(2+Lo-Lo\tau_{C_v})}{2(Lo+1-Lo\tau_{C_v})}, \quad (5.26)$$

where bw_{s_0} is the fractional stationary bandwidth without load.

According to Fig. 5.5, the dynamic fractional bandwidth is presented by the sum of the half stationary bandwidth, namely $0.5bw_{s_1}$ and $0.5bw'_{s_1}$, and frequency tunability, namely τ_f:

$$bw_d = \frac{bw_{s_0}(2+Lo)}{4(Lo+1)} + \frac{bw_{s_0}(2+Lo-Lo\tau_{C_v})}{4(Lo+1-Lo\tau_{C_v})} + \frac{Lo\tau_{C_v}}{2(Lo+1)}. \quad (5.27)$$

As illustrated in Fig. 5.8(b), the condition for enhancement of dynamic bandwidth is:

$$bw_d > bw_{s_0}. \quad (5.28)$$

The roots of the inequality are given as:

$$\tau_{c1} = \frac{1}{4Lo}\left[2 + bw_{s_0} + 2Lo + 2bw_{s_0}Lo \right.$$
$$\left. -\sqrt{-8bw_{s_0}Lo(2+Lo) + (2+bw_{s_0}+2Lo+2bw_{s_0}Lo)^2}\right] \quad (5.29)$$

$$\tau_{c2} = \frac{1}{4Lo}\left[2 + bw_{s_0} + 2Lo + 2bw_{s_0}Lo \right.$$
$$\left. +\sqrt{-8bw_{s_0}Lo(2+Lo) + (2+bw_{s_0}+2Lo+2bw_{s_0}Lo)^2}\right]. \quad (5.30)$$

As illustrated in Fig. 5.9, in the range of narrow stationary bandwidth the low root is above zero, while the high root is always beyond one which imposes actually no constrain on the required tunability.

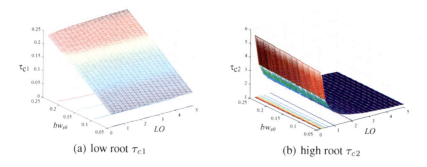

(a) low root τ_{c1}

(b) high root τ_{c2}

Fig. 5.9 Roots of inequality Eqn. 5.29

Therefore, the required tunability for the bandwidth enhancement is given by:

$$\tau_c > \tau_{c1} \,. \tag{5.31}$$

In Fig. 5.9(a) it suggests that, with narrow stationary bandwidth, the required minimal varactor tunability is linearly related to the stationary bandwidth. This can be further proved by the following approximation:

$$\frac{\tau_c}{bw_{s_0}} > \frac{(1-\tau_c)Lo^2 - 0.5Lo\tau_c + Lo}{Lo(Lo - Lo\tau_c + 1)} \approx 1 \,. \tag{5.32}$$

Thus, under the condition of narrow stationary bandwidth and low tunability, the minimal tunability is only required to be larger than the stationary bandwidth according to Fig. 5.10, where the approximation is noted as dotted line. The group of solid lines below the dotted line represents the influence of loading factor. The influence of loading factor is limited, while the deviation from the approximation increases at wider unloaded stationary bandwidth.

When the varactor tunability is larger than the stationary fractional bandwidth of the unloaded static antenna, i.e. $\tau_c > bw_{s_0}$, the dynamic fractional bandwidth of the tunable antenna is always increased. It provides a simple but referential criteria for both, the antenna design and the screening of tunable component technologies.

5.1.2 Nonlinear Model of Capacitively Loaded Planar Inverted-F Antenna

In capacitively loaded tunable antennas, the varactors exhibit a voltage-dependent capacitance under both DC and RF voltages. Especially when a varactor loaded

5.1 Tunable Multiband Antennas

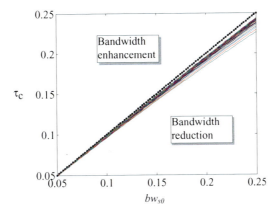

Fig. 5.10 Requirement on varactor tunability. The dynamic fractional bandwidth of the tunable antenna always increases, when the varactor tunability is larger than the stationary fractional bandwidth of the unloaded static antenna.

antenna is stimulated around its resonance, the varactor may generate a considerable harmonic current, which causes the harmonic radiation. Unless using a filter in serial connection with the varactor, which leads to a higher loss and a frequency detuning, the harmonic radiation can hardly be filtered out and hence turns to be critical. This section describes an efficient method to model the nonlinearity of tunable antennas. Different from analyses on the radiation of the fundamental resonance with small signal stimulation [3, 28, 109], harmonic radiation analyses take account of transient change of component parameters. They try to provide the far-field power level and radiation pattern at both fundamental and harmonic frequencies for specifically given input power. A full-wave method based on the finite-difference time-domain method (FDTD) method has been reported by Kalialakis and Hall in [51]. An analytical C-V dependency of the varactor is interpolated from discrete datasheet; a Gaussian pulse excites the varactor loaded antenna; the near-field is calculated by FDTD and then transformed to far-field, which yields the radiated power after a spatial integration. In the full-wave method, several dedicated analyses are necessary for varied varactor capacitances and input power levels, which are drastically time consuming for narrow band antennas with high Q-factor.

A fast method based on a wide-band equivalent-circuit model is developed and verified in this section. At a comparable accuracy to the full-wave analysis, this method enables direct spectral investigation of harmonic radiations from a varactor loaded antenna, and their dependence on input power level and varactor's DC bias voltage. Without loss of generality, a single-band planar inverted-F antenna (PIFA) encompassing a capacitively top-loaded short monopole and a shunted ground post as inductive matching is exemplified here. The antenna's internal capacitance, inductance and radiation form a parallel lossy resonator [71], which transforms the port impedance to the far-field wave impedance. By varying the geometries of the antenna patch, ground plane and the distance in between, the reactance can be chosen to maximize power radiation at given feeding frequency and port impedance. When a lumped capacitor is loaded around the open end of the patch, as illustrated in Fig. 5.11, the resonant frequency is lowered. By using a varactor instead, an electrically tunable PIFA can be realized.

A single-band tunable antenna is implemented within a typical hand-held volume. The design is optimized in the commercial full-wave solver CST Microwave Studio. The antenna patch and ground plane are produced on FR4 substrate of 1.5 mm thickness. An Infineon BBY52-03W silicon diode is embedded at the end of ground post, the capacitance of which changes from 2.6 pF to 1.1 pF under a 0 V to 4 V DC bias voltage. The DC bias is decoupled from RF path by the bias network in Fig. 5.13.

Table 5.2 Geometries of single-band tunable planar inverted-F antenna in Fig. 5.12, all in mm unit

l_g	w_g	l	w	h	d_{stub}	d_{pin}	Φ
80	44	18	37	7	2.8	9.7	1

The prototype is measured for its wide-band reflection coefficient. As shown in Fig. 5.14, the frequency where the antenna impedance is best matched to 50 Ω is consequently tuned from 0.868 GHz to 1.15 GHz. The instantaneous bandwidth of -6 dB reflection varies from 30 MHz to 64 MHz.

It is reasonable to assume that, except the varactor all the structures in the PIFA are independent to the input power level and not electrically tunable. In other words, these parts can be modeled as linear equivalent-circuit components and directly applied to large signal analyses. In detail, a two-port scattering matrix is obtained in the full-wave solver when the varactor is replaced by a 50 Ω port. In order to estimate the radiation power, a two-port equivalent circuit including the explicit radiation resistance is modeled by fitting its scattering matrix to the one from full-wave method. The nonlinear varactor model is finally loaded to the equivalent circuit, which enables large signal analysis.

The impedance characteristic of a single-band PIFA close to its resonance can be modeled as a parallel resonator [3, 28, 109]. However, such lumped resonator approximation is not valid for the harmonic analysis, since a part of the reactance comes from distributed parasitics and the impedance transformation across the patch, all of which are frequency-dependent. Additional components are introduced as shown in Fig. 5.15. The transmission line accounts for phase delay from the patch's end to the feeding point, i.e. the impedance change over the patch. It results in higher-order resonances, which is essential for the harmonic radiation analysis. In order to directly compare with measurements, the capacitance and quality factor of the varactor in the equivalent circuit are kept same as those in measurement. A transformer is used to represent the impedance discontinuity from the feed pin to the patch. The inductance of the pin L_{pin} grounded through varactor is calculated as in [59]:

5.1 Tunable Multiband Antennas

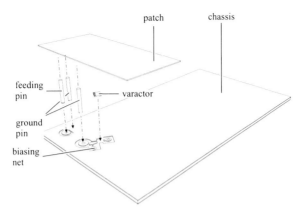

Fig. 5.11 Assembly perspective of the planar inverted-F antenna with a varactor load

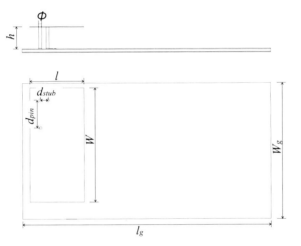

Fig. 5.12 Side view and top view of the tunable planar inverted-F antenna, all geometries are listed in Table 5.2

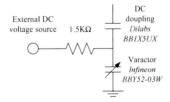

Fig. 5.13 Biasing network for the varactor

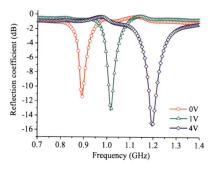

Fig. 5.14 Measurements of reflection coefficients of the diode loaded planar inverted-F antenna

$$L_{pin} = 2h[2.303\log(\frac{4h}{\Phi}) - 1 + \frac{\mu}{4} + \frac{\Phi}{2h}], \qquad (5.33)$$

where L_{pin} is the inductance in nH, h is the straight wire length and Φ is the diameter of the wire, both in cm. The self-inductance of the feeding pin and ground pin, namely L_{feed} and L_{stub} are obtained through parametric fitting of full-wave simulation results, as well as the transformation ratio n, the impedance Z_{TL}, length l_{TL} and effective relative permittivity ϵ_{TL} of the transmission line, and the radiation resistance R_{rad}.

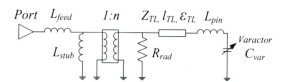

Fig. 5.15 Wide-band equivalent circuit of the single-band tunable planar inverted-F antenna

Since the parametric fitting takes only the linear parts between input port and the varactor into account, an additional 50 Ω port is used to replace the varactor in the calculation of two ports scattering matrix. The following cost function, i.e. E_{fit}, accounts for both reflection and transmission characteristics of the structure, when utilizing the two-port scattering-matrix is used:

$$E_{fit}(L_{feed}, L_{stub}, L_{pin}, n, R_{rad}, Z_{TL}, l_{TL}, \epsilon_{TL}) = \int_{f_L}^{f_H} \left\{ \sum_{i,j=1}^{2} \frac{\left|S_{ij}^{fw} - S_{ij}^{ec}\right|}{\left|S_{ij}^{fw}\right|} \right\} df, \qquad (5.34)$$

where the superscript fw refers to the scattering coefficients from full-wave method while the ec refers to those from equivalent-circuit method. Meanwhile all conductive loss is omitted for a more reliable estimation of the radiation resistance.

5.1 Tunable Multiband Antennas

The integration boundaries f_L and f_H are 0.5 GHz and 3 GHz respectively. The fitted parameters are listed in Table 5.3. A simple frequency-dependent interpolation model is used for the radiation resistance as in Table 5.4.

Table 5.3 Fitted parameters of equivalent circuit model in Fig. 5.15

L_{feed}	L_{stub}	L_{pin}	n	Z_{TL}	l_{TL}	ϵ_{TL}
3.28 nH	0.8 nH	3.9 nH	2.25	32.5 Ω	25.6 mm	3

Table 5.4 Interpolated frequency-dependent radiation resistance in Fig. 5.15

R_{rad}				
868 MHz	995 MHz	1.50 GHz	1.572 GHz	2.80 GHz
600 Ω	440 Ω	227.5 Ω	196 Ω	121 Ω

To ensure an accurate estimation of harmonic radiation levels from the equivalent circuit, two key performance indicators are verified: the port impedance and the radiation efficiency. The fitting error of input impedance is achieved less than 17 % around the fundamental and the first harmonic frequencies. As shown in Fig. 5.16, the model yields a reliable estimation of the input impedance over the frequency range from 0.5 GHz to 3 GHz.

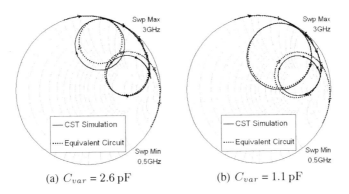

(a) $C_{var} = 2.6\,\text{pF}$ (b) $C_{var} = 1.1\,\text{pF}$

Fig. 5.16 Comparison of the input impedance between equivalent circuit method and full-wave method. Two cases with different C_{var} are given.

The radiation at the fundamental frequency is stimulated at the input port. Through the total efficiency η_{in}, one can figure out the power radiated at a given input power level:

$$\eta_{in} = \frac{P_{R_{rad}}}{P_{in}}, \tag{5.35}$$

where $P_{R_{rad}}$ refers to the power dissipation at the radiation resistance R_{rad}, and P_{in} is the input power. Compared to full-wave simulation with lossless conductors, the total efficiency differs less than 5 % within the varactor's tuning range, as in Table 5.5. Here the energy conversion from the fundamental to the harmonic resonant frequencies is already included.

Table 5.5 Comparison of simulated total efficiency η_{in} at fundamental frequency with -10 dBm input

C_{var} (pF)	1.1	1.8	2.6
η_{in} full-wave (%)	77	49	28
η_{in} equivalent circuit (%)	82	48	26

However, at harmonic frequencies, the varactor acts as a power source with complex impedance, while the input port is a 50 Ω resistive load. By a similar definition, the total efficiency η_{var} helps to understand how much the energy can be radiated at harmonics:

$$\eta_{var} = \frac{P_{R_{rad}}}{P_{var}}, \quad (5.36)$$

where P_{var} is the harmonic power generated by the varactor. In the comparison in Table 5.6, the varactor is represented as a 50 Ω port. Both full-wave method and equivalent-circuit method are carried out within the frequency range of harmonic resonances, where the alter one shows up to -2.1 dB relative difference.

Table 5.6 Comparison of simulated total efficiency η_{var} at harmonic frequencies from varactor port

f (GHz)	1.496	1.838	2.244
η_{var} full-wave (%)	46.7	39.6	23.5
η_{var} equivalent circuit (%)	35.5	24.8	14.6

As shown above, the proposed method requires only one full-wave analysis of the linear equivalent-circuit parts. The large-signal analyses can be then carried out for each varactor load on the equivalent circuit model. As a result, the whole process is faster even with a fine resolution of the varactor's capacitance and power levels.

It is the swing of RF voltage that transiently tunes the varactor and generates the harmonic current, which leads to radiation consequently. At a given input power level, the RF voltage across the varactor can be directly calculated with the equivalent circuit in Fig. 5.15. The varactor is initially assumed to be tunable only by DC voltage and constant under various RF voltage. The assumption yields the maximum amplitude of the RF voltage at the fundamental frequency. As shown in Fig. 5.17, a linear dependence of the RF voltage on the input power is lined, which is intuitive since all the parts are independent to input power level. As an example, when the antenna is fed with -10dBm, a 0.8V RF voltage is found on the varactor.

5.1 Tunable Multiband Antennas

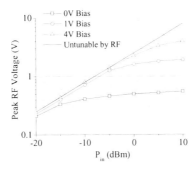

Fig. 5.17 Peak RF voltage over the varactor

For the diode loaded on the resonator, the considerable RF voltage can lead up to 25 % capacitance tuning. The modulation transiently shifts the resonant frequency and generates harmonics. Hence, when the RF voltage detunes the varactor, the reflection at the input port will increase due to frequency mismatch. Furthermore, the varactor excites harmonic current which affects the fundamental resonance, especially at lower DC bias voltage, where the varactor works partially in forward conduction region and exhibits strong nonlinearity. These effects introduce a saturation of the RF voltage on the varactor at the fundamental frequency. To characterize the effects, an equivalent-circuit model of the silicon diode up to 3 GHz in Fig. 5.18 is added to the PIFA model. The circuit and parameters are extracted from measurements and manufacturer's data. By using the harmonic-balance solver in AWR Microwave Office, several cases under different DC bias voltages are calculated and compared in Fig. 5.17. When increasing DC bias voltage, the nonlinearity of the varactor reduces, and the RF voltage saturates at higher input power. In other words, the power handling capability of the antenna increases with increased tuning voltage.

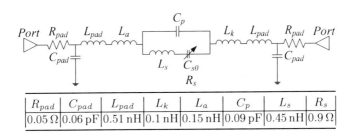

Fig. 5.18 Equivalent-circuit model of the Infineon varactor BBY52-03W

The spectrum of the power dissipation on R_{rad} is then obtained. In order to compare it with far-field measurements, the harmonic radiations are further weighted with additional free-space path losses:

$$HTRP_{2f_0} = P_{R_{rad2f_0}} - P_{R_{radf_0}} - 6\,dB$$
$$HTRP_{3f_0} = P_{R_{rad3f_0}} - P_{R_{radf_0}} - 12\,dB\,,$$
(5.37)

where $HTRP_{2f_0}$, $HTRP_{3f_0}$ are the total radiated harmonic power levels for the first and the second harmonic respectively [42]. The radiation spectrum at several bias voltages is shown in Fig. 5.19. With -10 dBm input power and 1 V DC bias voltage, where the varactor is still reversely biased but exhibits strong nonlinearity, the amplitude of the first harmonic at 2.01 GHz is -21.6 dBc, while for the second harmonic at 3.05 GHz it is -28.9 dBc. When the DC bias voltage increases, the harmonic radiation levels reduce together with a frequency shift.

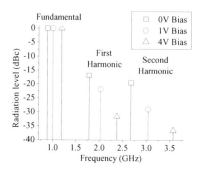

Fig. 5.19 Normalized radiation spectrum at -10 dBm input and DC bias voltage between 0 V and 4 V. Simulated by equivalent-circuit method.

In order to prove the above modeling method, the amplitude spectrum of far-field radiation is measured in an anechoic chamber. As illustrated in Fig. 5.20, the experiment setup includes HP8340A synthesized sweeper to stimulate the prototype at power level up to 10 dBm, EMCO3115 double ridged waveguide horn antenna as receiving reference, whose output is amplified first by Mini-circuit ZHL-1042J preamplifier, and then measured by HP8563A spectrum analyzer. The emission test setup covers a frequency range from 0.8 GHz to 2.9 GHz, which includes the fundamental and the first harmonic frequencies. Six planes are used to cut the antenna's far field pattern at 30° interval, while the rotation step within each plane is 15°. H- and E-plane polarizations are measured independently and then combined for the total radiated power (TRP). The directivity, total efficiency and gain are calculated using the three dimensional pattern integration method.

In Fig. 5.21, the normalized amplitudes of the fundamental and the first harmonic radiation on H-plane are presented in logarithmic scale. Due to the resonant frequency tuning and the harmonic excitation, the directivity at fundamental frequency changes from 2.6 to 2.4 when DC bias voltage increases from 1 V to 4 V. When fed with -10 dBm, the first harmonic radiation reduces from -20.8 dBc to -32.6 dBc due to the increase of the bias voltage. In comparison, the measured TRP of the first harmonic is between -21.6 dBc and -35.6 dBc. Beside the measurement uncertainties, the errors are mainly introduced by the different total efficiency shown in Table 5.6.

5.1 Tunable Multiband Antennas

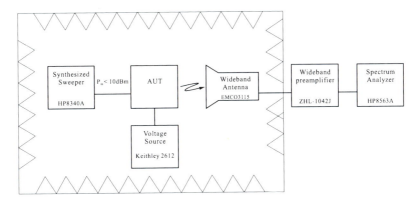

Fig. 5.20 Emission testing setup for far-field harmonic radiation

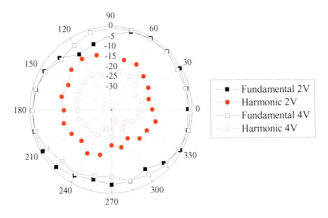

Fig. 5.21 Measurement of normalized radiation pattern on H-plane at the fundamental and the first harmonic frequencies. Two cases with 2 V and 4 V bias voltage are compared.

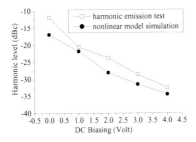

Fig. 5.22 Dependence of the first harmonic radiation level on DC bias voltage

The dependence of the harmonic radiation level on the DC bias voltage is given in Fig. 5.22. The input power is fixed at -10 dBm, which yields 0.8 V amplitude of the RF voltage on the varactor. At DC bias voltage lower than 0.8 V, the varactor is driven partially to the forward conduction region, where the PIFA is off resonance. This switching-like tuning results in a strong harmonic excitation. However, the varactor's forward conduction is not exactly modeled by the equivalent circuit; as a result the difference to the measurement is larger around 0 V biasing. Beyond this threshold, the antenna's linearity increases together with DC bias voltage. The trend also implies that the linearity of tunable PIFA can be improved simply by using varactors working at higher DC bias voltage.

5.1.3 Suppression of Harmonic Radiation by BST Thick-Film Varactors

Compared to diodes, the ferroelectric varactors built on BST thick-film exhibit symmetrical capacitance-voltage (C-V) dependence without forward conduction, low hysteresis and high break-down electric field strength. When implemented in tunable antennas, the varactors show advantages in improving the antennas' linearity. In this section, analyses and experiments are taken to evaluate the efficiency of the suppression of harmonic radiation by using such varactors.

The single-band tunable PIFA, as in Fig. 5.11, is taken to evaluate the RF voltage on varactor and the resulted radiated power spectrum. Since all the geometries are identical to those used in chapter 5.1.2, the suppression can be directly found. Instead of the diode in Fig. 5.13, a pair of ferroelectric varactors is applied as in Fig. 5.23, which are fabricated in the form of IDCs. They are of identical capacitance and in a serial connection. The DC biasing voltage is fed at the center of the pair. The configuration requires a resistor for DC-RF decoupling. The varactors are built on a standalone module, and connected through silver epoxy glue to the contact pads on PCB.

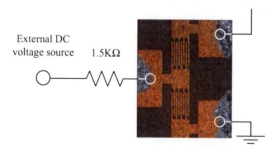

Fig. 5.23 The ferroelectric varactor pair implemented in the tunable planar inverted-F antenna

5.1 Tunable Multiband Antennas

A simulated comparison between using the BST varactor and the diode is shown in Fig. 5.24. The C-V dependence of the BST thick-film varactor is modeled by using a 15-order polynomial, which is valid between -100 V and 100 V. A 0.7 Ω serial resistor is used to model its loss. The model of the silicon diode is extracted from the measurement as in Fig. 5.18. Since the models of the BST varactor and the diode have limitations on the voltage range, the simulations are only valid up to 10 dBm and 30 dBm input, respectively. The comparison clearly shows two determining effects of the varactor's C-V behavior. First, the varactor is detuned by the considerable RF voltage at high input power level; the stimulation of harmonic currents weakens the fundamental resonance, and could reduce the radiation efficiency. Second, large RF voltage could drive the diodes to the forward conduction region. The voltage at fundamental frequency is then prone to saturate, resulting in an extraordinary harmonic excitation. For BST varactors, the first factor comes to be critical only at a power level much higher than that with diodes, since their bias electrostatic field strength is so high that the RF signal only slightly detunes the varactors. The second factor is not applicable in BST varactor, which has no forward conduction in the operating voltage range and a very high break-down field strength.

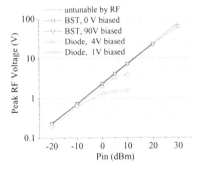

Fig. 5.24 The peak RF voltage over the varactor at fundamental frequency, which are compared between using BST varactor and silicon hyperabrupt diode. All are simulated at 1 GHz.

The radiation spectrum with -10 dBm input power is then calculated at several bias voltages as shown in Fig. 5.25. Here only the best input matched frequencies are considered. Besides the above mentioned models of the BST varactor and the diode, additional models are introduced, namely *PINDEV* and *ThinBST*. In *PINDEV*, the C-V dependence is scaled from that of the diode, i.e. it changes capacitance from 2.6 pF to 1.1 pF under 0 V to 90 V, while the forward voltage keeps. Hence, the influence of the diode's low tuning voltage is mitigated. It helps to reveal the effect of the forward conduction. The other model *ThinBST* stands for a BST thin-film varactor as in chapter 3.2. The symmetric C-V dependence with 28.5 % tunability is same as that of BST thick-film varactors, while the bias voltage is reduced from 90 V to 6 V. It is clear that, by replacing the diode with the BST thick-film varactors, the first and the second harmonic radiations are efficiently suppressed by 27 dB to 65 dB. Moreover, the suppression is stable over the tuning range. The scaled diode shows comparable if not better suppression at 90 V biasing, which ensures no forward conduction and tunes the varactor to a weaker C-V dependent region. But at 1 V it is still partially forward conducting, and excites up to -40 dBc harmonic radiation. The BST thin-film varactor is expected to show an intermediate harmonic radiation between those of diode and BST thick-film. Because of no forward conduction, it exhibits a small fluctuation of the harmonic radiation across the whole tuning range.

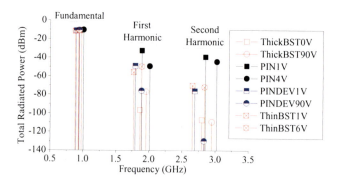

Fig. 5.25 Normalized radiation spectrum at -10 dBm input and varied DC bias voltages

In the prototype, the BST varactors are produced on a 4.1 μm thick layer of $Ba_{0.6}Sr_{0.4}TiO_3$. The BST layer was screen-printed and sintered on top of a 640 μm Al_2O_3 ceramic substrate. At room temperature, the BST layer shows a loss tangent of 0.009 and relative permittivity of 410 at 1 GHz. The planar IDC is patterned on top of the layer using the lithographic method. The fabrication procedure is described in section 3.1.2. The electrodes here are 2.5 μm thick at a gap width of 8.1 μm. The capacitance of each varactor is measured from 2.84 pF to 2.09 pF under 0 V to 90 V DC bias voltage, i.e. a tunability of 28.5 %. The Q-factor is above 80 at 1 GHz under room temperature. Meanwhile the DC leakage current is below 2.7 nA. As depicted in Fig. 5.26, the frequency where the antenna impedance is best matched to 50 Ω can be continuously tuned from 0.99 GHz to 1.12 GHz. The measured radiation efficiency varies from 37 % to 61 %. The -6 dB instantaneous bandwidth varies between 21.8 MHz and 35 MHz.

The radiation spectrum in far-field of the prototype is then measured in an anechoic chamber, using the setup as in Fig. 5.20. The frequency range is from 0.8 GHz to 2.9 GHz, covering the fundamental and the first harmonic frequencies. Since the prototype's harmonic radiation is close to the setup's sensitivity, several methods are taken to secure reliable measurement, including reducing the path distance, limiting to 90 V biasing voltage and averaging measurements. Fig. 5.27 shows normalized pattern of the fundamental and the first harmonic radiation on H-plane. The BST varactors are biased at 90 V to tune the PIFA to a center frequency of 1.12 GHz. The results are compared with the radiation of the diode loaded prototype in section 5.1.2. A 1 V biasing voltage tunes the diode loaded prototype to 1.015 GHz. The two prototypes are fed identically with -10 dBm input power, which excite the strongest harmonic radiation in their own cases. The TRP is then calculated

5.1 Tunable Multiband Antennas 79

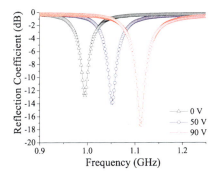

Fig. 5.26 Measured reflection coefficients of the BST thick-film varactor loaded planar inverted-F antenna

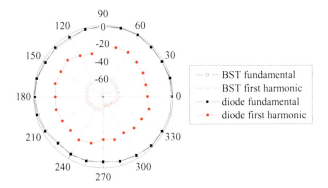

Fig. 5.27 Measurement of normalized radiation pattern on H-plane at the fundamental and the first harmonic frequencies, by using BST thick-film varactor and diode. It shows a considerable suppression by the BST varactor.

by three-dimensional pattern integration. A 35.7 dB relative suppression of TRP at the first harmonic is observed, which is 44 dB in the simulation. Besides the measurement uncertainties, the diode's model in forward region mainly introduces the error.

5.1.4 Independently Tunable Multiband Slot Antenna

As shown above, the limited frequency tunability constrains the frequency coverage of the single-band antennas to only neighboring services. In order to cope with applications with scattered frequency bands, multiband tunable antennas are desirable. On one hand, they provide a flexibility in allocating dedicated bands for widely separated services. On the other hand, the multiple bands could operate simultaneously while reducing the occasion of frequency tuning during operation. Furthermore, the frequency tuning of each band is desired to be independent to those of other bands, which allows convenient coverage control for individual services. This section is to propose a novel design of a dual-band tunable antenna. The two bands are independently controlled by corresponding varactors, which further enables adaptive mitigations of the environmental detuning, manufacture tolerance and interference in the two bands independently. In addition, the antenna can be easily accommodated on a chassis with stringent footprint. The design method of single-band tunable element is presented first, followed by the dual-band concept and measurements.

The basic element for the single-band tunable antenna is a capacitively loaded open-ended monopole slot antenna, as in Fig. 5.28(a). The symmetrical voltage distribution of the fundamental mode about the longitudinal center of the $\lambda/2$ dipole slot is utilized to reduce the slot length to $\lambda/4$ [53], where λ denotes the wavelength in the slot. A varactor is loaded across the aperture at a distance l_0 from the open end. The overall slot length l determines the upper boundary of the achievable frequency range at the fundamental resonance. By varying the capacitance C and the position of the varactor, the resonance can be tuned to a lower frequency, i.e. the antenna can be further minimized.

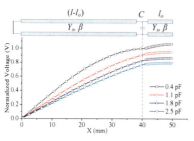 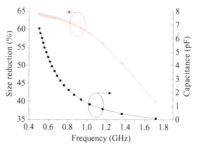

(a) Voltage distribution over a slot aperture.

(b) Frequency tuning by varying capacitance, and the corresponding size reduction.

Fig. 5.28 The tuning of resonant frequency by altering the slot current distribution with a loading varactor

With the depicted geometries, the resonant frequency ω is obtained by solving the transverse resonance condition:

$$-tan\left[\beta\left(l-l_0\right)\right] - ctan\left(\beta l_0\right) + \frac{\omega C}{Y_s} \cdot ctan\left(\beta l_0\right) \cdot tan\left[\beta\left(l-l_0\right)\right] = 0 \ . \quad (5.38)$$

The frequency-dependent propagation constant β and the characteristic admittance Y_s of the slot line are approximated in [47]. In Fig. 5.28(b), the needed capacitance is shown for the target frequency. In addition, the reduction of overall slot length compared to the length of the varactor loaded $\lambda/2$ dipole slot in [7] is also shown.

To reduce the occupation on the chassis, the slot is located at the edge and folded vertically to utilize the height dimension of the device's volume. Furthermore, the antenna is preferable to be simply made of metal sheet, so the feeding pin is designed in coplanar waveguide (CPW) form and can connect directly to a coaxial

5.1 Tunable Multiband Antennas

cable. In order to match the slot's high input impedance, the feeding is placed close to the shunted end. The architecture of the device is shown in Fig. 5.29(a), where a straight open-ended slot is formed between a strip line and a chassis build from a 1.5 mm-thick single sided FR4 board with a relative permittivity of 2.5. The reflection coefficients from a full-wave simulation by CST Microwave Studio is shown in Fig. 5.29(b). It is found that the transmission line model is referential in predicting resonant frequency. In other words, the differential current mode is dominant around the slot, which constrains the RF current in the vicinity of the slot and in return ensures the usage of the small chassis.

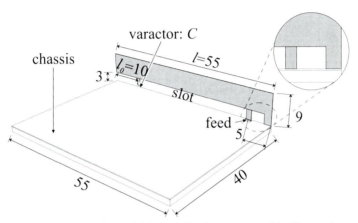

(a) Structure of a vertically folded tunable slot antenna, with offset coplanar waveguide feeding.

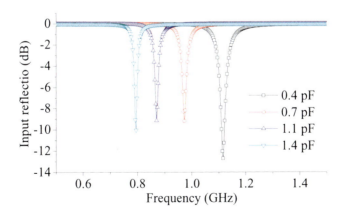

(b) Simulated tuning of the single-band tunable slot antenna, with varied capacitance.

Fig. 5.29 Single-band tunable slot antenna

By combining two above mentioned tunable slots at different frequencies, a dual-band prototype is realized as shown in Fig. 5.30. The two slot resonators are placed at neighboring edges of the chassis. They share a single feed and a single shunted inductor, which terminates both slots via a small rectangular path, see also in the zoom-in view. The feed's offset from the shunted end in each of the slots, namely l_{pl} and l_{ph} for the low band and the high band, respectively, can be independently optimized by varying the patch's geometry. The geometries, especially the positions of the varactors, are optimized so that the resonators can be tuned independently.

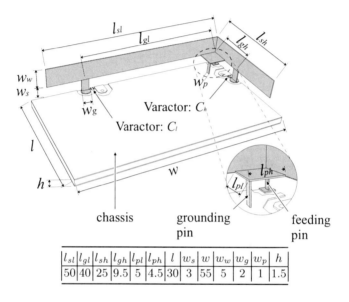

l_{sl}	l_{gl}	l_{sh}	l_{gh}	l_{pl}	l_{ph}	l	w_s	w	w_w	w_g	w_p	h
50	40	25	9.5	5	4.5	30	3	55	5	2	1	1.5

Fig. 5.30 Perspective of the dual-band tunable slot antenna. The overall dimension is 55 mm × 30 mm × 10 mm. All geometries are in mm.

A prototype is then fabricated, as shown in Fig. 5.31. The overall dimension of the device is 55 mm × 30 mm × 10 mm. The antenna is made of a 0.1 mm-thick copper sheet. For each band, two Siemens BBY52-03W high Q hyperabrupt diodes as in section 5.1.2 are connected in an anti-serial manner. Their capacitance varies from 2.5 pF to 0.8 pF under the DC voltage from 0.5 V to 6 V. The biasing voltage below 0.5 V is intentionally avoided to mitigate the harmonic excitation. The DC voltage feeds through the middle between the varactors, and are resistively decoupled from the RF signal with 40 kΩ resistors. The two varactor pairs are noted as C_l for the low band slot and C_h for the high band slot.

The input reflection of the prototype has been measured when applying different bias voltages. As in Fig. 5.32(b), when the bias voltage V_l for C_l is fixed and only a varied V_h is applied to C_h, the central frequency of high band resonance is tuned from 1.51 GHz to 2.09 GHz. With a -6 dB reflection criteria, the tunable band equivalently covers the range from 1.45 GHz to 2.25 GHz. At the lower

5.1 Tunable Multiband Antennas

Fig. 5.31 The realized prototype with RF feed and DC control cables

end where V_h is 0.5 V, the antenna is not well matched, which results in an instantaneous bandwidth of 91 MHz. For higher voltages with V_h greater than 1 V, the instantaneous bandwidth is above 150 MHz, which is sufficient to cover single services like digital cellular system (DCS), personal communications service (PCS) and barely UMTS. With the same -6 dB equivalent bandwidth criteria, the low band covers the range from 705 MHz to 957 MHz, as in Fig. 5.32(a). However, the low band's instantaneous bandwidth is in the range from 15 MHz to 44 MHz, which is insufficient to cover simultaneously the down- and uplink channels in semi-duplexing services as GSM850 and EGSM900. One solution is to tune dynamically the resonance to hop between the transmission and receiving channels. The measured total efficiency varies between 41 % and 68 %. Especially at the bands' lower boundaries, the heavy capacitive load leads to additional loss on the varactors. Hence, a high Q-factor of the varactors is essential to improve the efficiency.

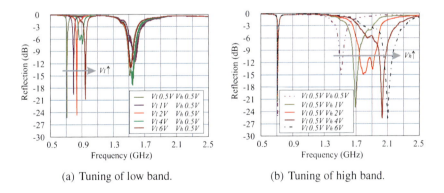

Fig. 5.32 Measured independent tuning of the two bands

5.1.5 Integrated Duplex Ceramic Antenna

In the above sections, efforts are taken to increase the dynamic coverage of a single-band resonance, and to introduce a second resonance to cover multiple services. However, the limited instantaneous bandwidth of the individual-bands is hardly enough to cover simultaneously the downlink and uplink channels in FDD applications. This section presents a novel antenna design with integrated ferroelectric varactors. It provides not only a tunable center operating frequency, but also a variable frequency separation between downlink and uplink channels. The antenna consists of two adjacent slots, both of which are loaded with ferroelectric varactors. The varactors are biased through integrated resistive strip lines. In a prototype, the antenna is mounted at the edge of a small chassis which is evaluated for reflection and far-field measurements.

The configuration of the antenna is depicted in Fig. 5.33. It consists of a ceramic antenna, microstrip feed and chassis. The tunable antenna is built on a piece of ferroelectric ceramic substrate. The microstrip feed is built on a single sided 0.8 mm-thick FR4 substrate. The antenna is mounted at the chassis' edge. The feed port, ground and bias are connected through spring connectors to the chassis. On the antenna, across each slot there is an interdigital capacitor (IDC) pair close to the open end. The varactor pairs namely C_1 and C_2 load at the bottom and the upper slot, respectively. Since they are built on the BST thick-film, the capacitance is varied by applying external electrostatic field across the gaps between digits, thus tuning the slots. In a similar topology as in section 5.1.4, the IDC pair is biased at the middle through a highly resistive line. The integrated topology decouples RF signal and external electrostatic bias, with affordable compromise in the overall tunability and Q-factor [83]. The two varactor pairs are independently tunable. Therefore, by tuning their capacitances synchronously, the center frequency F_C at the middle of the two channels is tuned. And by varying the ratio between C_1 and C_2, the channel separation B_{SP} is altered.

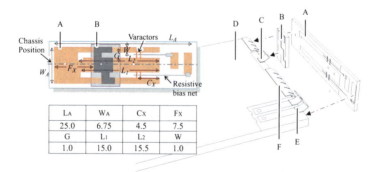

Fig. 5.33 Perspective view of A: planar ceramic antenna, B: microstrip feed, C: feed connector, D: chassis, E: DC bias connector, F: ground pins. All Geometries are in mm unit.

5.1 Tunable Multiband Antennas

With the above mentioned parameters, a prototype is realized as depicted in Fig. 5.34. The antenna is fabricated on a 650 µm-thick Al_2O_3 substrate with 3.5 µm BST thick-film screen printed on top. The BST-layer exhibits a relative permittivity of 440 at 2 GHz and room-temperature, with a loss tangent of 0.011. A thin chromium and gold seed layer is evaporated above. Then the IDCs are realized with 3.8 µm-thick plated gold electrodes. Additional 50 µm-thick gold is plated on strip lines. 40 µm-wide and approximately 30 nm-thick bias lines are etched on the chromium seed layer, which show 4 kΩ/mm resistivity per line length. The antenna and the feed are edge mounted using a holder made from thin Plexiglas. To ease the manufacturing, spring loaded connectors are utilized instead of springs.

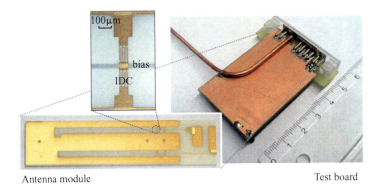

Fig. 5.34 The realized prototype and the varactor pair with bias network

The untuned capacitance of each varactor pair is 1.54 pF. With 100 V bias voltage across the 6 µm gap, 46 % tunability and Q-factor above 65 are achieved in the frequency range. The antenna's reflection coefficients are measured during tuning as shown in Fig. 5.35. The slots are controlled in two modes. The two slots vary their resonant frequency together when the two varactors are tuned proportionally. Therefore, the central frequency F_C shifts while the separation bandwidth B_{SP} keeps almost constant as shown in Fig. 5.35(a). In the second mode in Fig. 5.35(b), the ratio between the two varactors' capacitance varies, which leads to a considerable change of B_{SP} but a negligible shift of F_C. The independent tuning of F_C and B_{SP} provides an efficient way to cover different FDD services in the frequency range.

As analyzed in section 5.1.1, the constrains on the frequency tunability mainly come from the loading factor and the varactor's tunability. The loading factor is determined by both, the position and the untuned capacitance of the varactor, which are fixed here. The limited varactor tunability τ reduces the range when a certain ratio between C_1 and C_2 is required. Such constrain is illustrated in Fig. 5.36(a). Within the range from $C_{2F_{min}}$ to $C_{1F_{max}}$, the C_2/C_1 ratio between 1.5 and 1.1 is fully achievable which guarantees the attainable variation of separation bandwidth. Outside the range, the center frequency is further tunable, however the variation of B_{SP} is reduced. The measured tuning range is summarized in Fig. 5.36(b), the F_C is

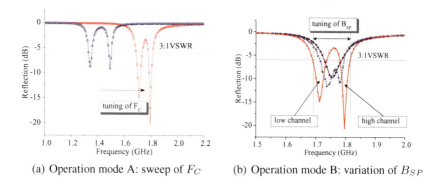

(a) Operation mode A: sweep of F_C

(b) Operation mode B: variation of B_{SP}

Fig. 5.35 Reflection coefficient measurement of prototype. Both the center frequency F_C and the separation bandwidth B_{SP} can be controlled.

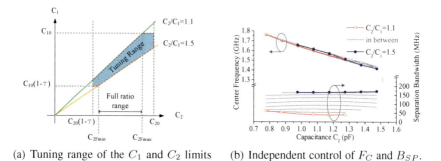

(a) Tuning range of the C_1 and C_2 limits the coverage of F_C and B_{SP}.

(b) Independent control of F_C and B_{SP}.

Fig. 5.36 Independent tuning of center frequency and separation bandwidth, and the constrains from varactor tunability

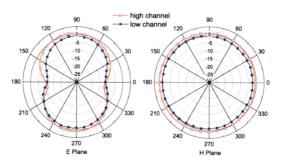

Fig. 5.37 Radiation pattern at center frequency $F_C = 1.76$ GHz and separation bandwidth $B_{SP} = 181$ MHz. The pattern of the high and low channels are close to each other, which allows a reliable frequency division duplex link.

tunable from 1.47 GHz to 1.76 GHz. The maximal distance between the low uplink channel and the high downlink channel, which is defined at -6 dB matching level, is variable from 38 MHz to 181 MHz. However the controlling of the two parameters are well decoupled.

Radiation measurement shows a bidirectional E-plane pattern and an omnidirectional H-plane pattern as depicted in Fig. 5.37. The pattern alters slightly between the two adjacent channels. The measured total efficiency is higher than 35 %.

5.2 Tunable Single-Band Impedance Matching Network for Antennas

Impedance matching networks are widely implemented between antennas and frontends to maximize the total efficiency of the antennas. The deviation of antenna impedance may rise from engineering tolerance, which reduces the yield ratio if the reflection exceeds a threshold across the target operating frequency range. Furthermore, as measured in Fig. 5.38, compact antennas are subject to environmental surrounding which functions as reactive or resistive load on the resonant antennas, and therefore, alters the impedance. In both cases, electrically tunable impedance matching networks are desirable to compensate the inevitable impedance mismatch. In recent years, developments in materials and processing techniques of the ferroelectric material like BST have enabled continuously tunable matching networks (TMNs) with promising performance [19, 84]. Different processing techniques i.e. thin- and thick-films, as well as varied matching network topologies, e.g. the Π- and T-structures, have been investigated [85]. The tunable matching network (TMN) based on BST thick-film varactors have proven to improve the impedance matching, while keeping a high power handling capability and working at a negligible current

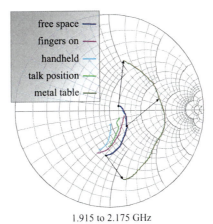

Fig. 5.38 Measurement of the environmental influence on the impedance of a handheld antenna. The smith chart is normalized to 50 Ω.

consumption. Additionally, in order to maintain the performance under changing conditions, an adaptive control is indispensable.

A simple TMN consists of lossy varactors and static inductors. This section is to identify the theoretical constrains on its performance at single frequency. It later analyzes the bandwidth enhancement for frequency dependent loads. The design, realization and integration technologies of demonstrative prototypes are then given in detail, as well as the means of efficient adaptive control.

5.2.1 Theoretical Analysis: Single-Band Design Guidelines

A typical TMN is designed to enlarge the impedance matching range through tuning its varactors. Two simple and functional types of single-band TMN are in Π- and T-topologies as shown in Fig. 5.39(a) and 5.39(b), respectively. They bridge the source and the load with two varactors, namely C_1, C_2, and one inductor L. The varactors are biased by external DC voltage source through DC-RF decoupling peripherals. The DC biasing network is not shown here.

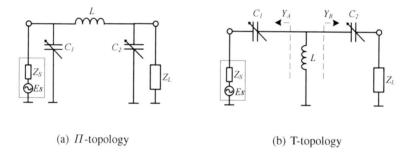

(a) Π-topology   (b) T-topology

Fig. 5.39 Representative topologies of single-band lossless tunable matching networks

With lossless components, both networks allow complex conjugate matching simultaneously at the input and the output ports to the source and the load impedance respectively. As illustrated in Fig. 5.40, a T-topology TMN covers the highlighted range where the complex source and load impedances are conjugate matched. When the source has negligible reactance, which is typical for prematched frontends, the minimal reflection at TMN's input port and the maximal power delivered to the load can be achieved at the same time.

In the following, a novel explicit design flow is proposed as depicted in Fig. 5.41. It starts with single-band lossless topologies. For T-networks, the specific close-form solutions of the component parameters are derived at single frequency point, which is then extended to consider frequency dependent load in section 5.2.3. With static inductors in the networks, the influence on network Q factor is then analyzed, which introduces the fundamental constrain on the adaptive control in section 5.2.5. Finally, the shift of the optimal operation point due to the loss in reactive components is quantified, which is later verified by experiments.

5.2 Tunable Single-Band Impedance Matching Network for Antennas

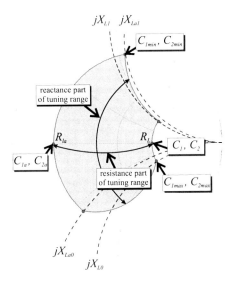

Fig. 5.40 Design parameters and the conjugate matching range of a typical T-topology tunable matching network

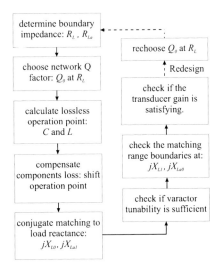

Fig. 5.41 The proposed design approach of single-band tunable matching network

Tunable Matching Network with Static Inductance

For given source and load impedances, an analytical solution of the optimal lossless capacitors and inductors has been given in [112] for Π-networks. It is to be extended to T-networks here. The following analysis is for one single frequency point, therefore, the frequency dependence of the network, the source and the load are neglected here. However, for wide band operations, these components are inevitably frequency dependent. The resulted performance shift is discussed later in section 5.2.2.

In the lossless T-network as in Fig. 5.39(b), the complex load can be matched in the imaginary and the real parts respectively. As illustrated in Fig. 5.42(a), when a reactance conjugate to the load reactance is inserted between the TMN's output and the load, the load turns to be only resistive. In detail if the load is inductive as in Fig. 5.42(b), then a capacitor as in Eqn. 5.39 is implemented, where ω is the angular operating frequency and L_L is the load inductance. Otherwise, when the load is capacitive as in Fig. 5.42(c), a negative capacitor as in Eqn. 5.40 is implemented and in practice equals to a reduction of output capacitance, where C_L is the load capacitance. The former method is named resonance method and the latter is absorption method [97].

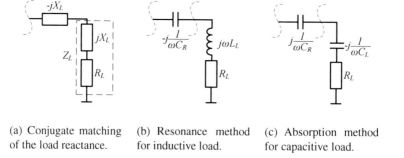

(a) Conjugate matching of the load reactance. (b) Resonance method for inductive load. (c) Absorption method for capacitive load.

Fig. 5.42 Conjugate matching of the load reactance, using the resonance and absorption methods

$$C_R = \frac{1}{\omega^2 L_L} \quad (5.39)$$

$$C_R = -C_L \quad (5.40)$$

Afterwards, there are only resistive R_S and R_L of the source and the load respectively. At the single frequency point, the source and load impedance can be represented by conductance:

5.2 Tunable Single-Band Impedance Matching Network for Antennas

$$G_S = \frac{1}{R_S}$$
$$G_L = \frac{1}{R_L}, \quad (5.41)$$

where the R_S and R_L denote the normalized resistance of the source and the load respectively. Therefore the Q factor at the input port and the output port, namely Q_S and Q_L respectively, are defined as:

$$Q_S \equiv \frac{G_S}{B_{C_1}}$$
$$Q_L \equiv \frac{G_L}{B_{C_2}}, \quad (5.42)$$

where B_{C_1} and B_{C_2} are the susceptance of C_1 and C_2, respectively.

Therefore, the admittance Y_A at left of the inductor as shown in Fig. 5.39(b) is given as:

$$Y_A = \frac{1}{1/G_S + 1/(j\omega C_1)} = \frac{G_S + jQ_S G_S}{1 + Q_S^2}. \quad (5.43)$$

The effective conductance G_A as the real part of Y_A is represented by the source conductance and the source Q factor. In the same way, such relationship is found at the output port:

$$G_A = \frac{G_S}{1 + Q_S^2} \text{ and } B_A = \frac{Q_S G_S}{1 + Q_S^2} = G_A Q_S$$
$$G_B = \frac{G_L}{1 + Q_L^2} \text{ and } B_B = \frac{Q_L G_L}{1 + Q_L^2} = G_B Q_L. \quad (5.44)$$

The conjugate matching condition at the junction of the inductor is as following:

$$Y_A^* = Y_B - jY_L, \quad (5.45)$$

where the Y_L is the admittance of the inductor L.

The condition is rewritten in the real and the imaginary parts:

$$G_A = G_B$$
$$-jB_A = jB_B - jY_L. \quad (5.46)$$

Here, a network Q factor namely Q_0 can be defined as the quotient between the inductor susceptance and the total conductances:

$$Q_0 \equiv \frac{Y_L}{G_A + G_B} = \frac{G_A Q_S + G_B Q_L}{2G_A} = \frac{Q_S + Q_L}{2}. \quad (5.47)$$

By substituting G_A, G_B, B_A and B_B with the effective conductance and the network Q factor in Eqn. 5.47 and the real part of 5.46, the source and load Q factors are represented as:

$$Q_S = \frac{2Q_0 G_S - \sqrt{4Q_0^2 G_S G_L - (G_S - G_L)^2}}{G_S - G_L}$$

$$Q_L = \frac{2Q_0 G_L - \sqrt{4Q_0^2 G_S G_L - (G_S - G_L)^2}}{G_L - G_S}. \quad (5.48)$$

Then applying the above Q_S and Q_L in the imaginary part of Eqn. 5.46, the reactance of C_1 and C_2 of T-network can be found as:

$$\omega_m C_1 = \frac{(G_S - G_L) G_S}{2Q_0 G_S - \sqrt{4Q_0^2 G_L G_S - (G_S - G_L)^2}}$$

$$\omega_m C_2 = \frac{(G_L - G_S) G_L}{2Q_0 G_L - \sqrt{4Q_0^2 G_L G_S - (G_S - G_L)^2}}, \quad (5.49)$$

where ω_m is the angular frequency.

For chosen network quality factor Q_0, L is further given:

$$\omega_m L = \frac{1}{(G_S Q_S)/(1+Q_S^2) + (G_S Q_L)/(1+Q_S^2)} = \frac{1+Q_S^2}{2Q_0 G_S}$$

$$= \frac{2Q_0(G_L + G_S) - 2\sqrt{4Q_0^2 G_L G_S - (G_S - G_L)^2}}{(G_L - G_S)^2}. \quad (5.50)$$

It is concluded that for a T-network the both capacitors and the inductor are determined by the network Q factor, the source and the load conductance. Once the Q factor is chosen, the network parameters namely the operation points are given in analytic close-form. Table. 5.7 shows an example where for a chosen Q_0 of 4 at 2 GHz a T-topology TMN is designed for a 150.5 Ω and a 16.6 Ω load, respectively.

Table 5.7 Exemplary network designs with $Q_0 = 4$

Load(Ω)	C_1(pF)	C_2(pF)	L(nH)
150.5	0.3091	0.1854	13.68
16.6	0.5580	0.9305	4.54

As shown above, when using such network to cover a range of varied load impedance, the ideal operation points of the capacitance and the inductance alter along with the load. However, since variable low loss lumped inductors are not available, fixed inductors must be used. Once the network Q factor Q_0 and L are

5.2 Tunable Single-Band Impedance Matching Network for Antennas

chosen for a specific load G_L, the network Q factor Q_{0a} for a varied load G_{La} can not be chosen freely but becomes a dependent variable which is determined by:

$$Q_{0a} = \frac{k_p + \sqrt{k_p^2 - 4\frac{1}{g_{La}^2 \omega_m^2 L^2} - k_n^2}}{2\frac{1}{G_{La}\omega_m L}}, \quad (5.51)$$

where

$$k_p = (k_a + 1)$$
$$k_n = (k_a - 1)$$
$$k_a = G_S/G_{La}$$
$$G_{La} = 1/R_{La}. \quad (5.52)$$

The matching to G_{La} is then achieved by tuning only the varactors. The required capacitance is obtained then:

$$\omega_m C_{1a} = \frac{(G_S - G_{La})G_S}{2Q_{0a}G_S - \sqrt{4Q_{0a}^2 G_{La} G_S - (G_S - G_{La})^2}}$$

$$\omega_m C_{2a} = \frac{(G_{La} - G_S)G_{La}}{2Q_{0a}G_{La} - \sqrt{4Q_{0a}^2 G_{La} G_S - (G_S - G_{La})^2}}, \quad (5.53)$$

which takes the drifted network Q factor and new load into account.

It is shown in Eqn. 5.51 that when L is fixed, the network Q factor Q_{0a} turns to depend on the load resistance. In T-network, the Q_{0a} increases with the decrease of load resistance, while in Π-network it increases with the increase of load resistance. Such dependence of Q_{0a} over a wide range of load impedance is exemplified in Fig. 5.43 on impedance Smith charts. When the Q_0 and L are designed, the Q_{0a} for another load G_{La} is subject to this determinative dependence. The dependence is observed on both Π and T topologies. The dependence of the Q_{0a} is mirrored between the two topologies. The initial Q_0 of T networks shall be chosen on the higher impedance boundary, which ensures the existence of Q_{0a} on the lower impedance side. For Π networks, one shall start from the lower impedance side.

Following the example in Table. 5.7, the T-topology TMN is designed with a static inductor as in Table. 5.8, for the same $Q_0 = 4$ at 2 GHz for a 150.5 Ω load. One shall notice that, when it is tuned to match 16.6 Ω, the Q_{0a} raises to 12.8, and the capacitors' operation points shift accordingly.

Considering the imaginary part by applying Eqn. 5.39 and 5.40 in Eqn. 5.49, the total capacitance of the both varactors is found as following. The reactance is matched by tuning the C_2 to C_{2T}, while C_1 is only responsible for the real part matching:

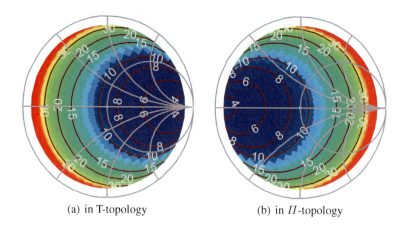

Fig. 5.43 Impedance dependence of Q_{0a} in T and Π topologies

Table 5.8 Exemplary network designs with a static inductor

Load(Ω)	C_1(pF)	C_2(pF)	L(nH)
150.5	0.3091	0.1854	13.68
16.6	0.1708	0.2951	13.68

$$C_{2T} = \frac{1}{\omega\left(1/\omega C_2 - X_L\right)}$$
$$C_{1T} = C_1 \,. \tag{5.54}$$

For the loads $R_{La} + jX_{La1}$ and $R_L + jX_{L0}$, the margins of C_2 are:

$$C_{2max} = max\,(C_{2T})$$
$$= \frac{1}{1/C_2 - \omega X_{L0}}$$
$$C_{2min} = min\,(C_{2T})$$
$$= \frac{1}{1/C_{2a} - \omega X_{La1}} \,. \tag{5.55}$$

The tunability of C_2 is:

$$\tau_{C_2} \equiv \frac{C_{2max} - C_{2min}}{C_{2max}} \,. \tag{5.56}$$

The margins of C_1 are defined by resistance range R_L and R_{La}, which are then given by:

$$C_{1max} = C_1$$
$$C_{1min} = C_{1a} \,. \tag{5.57}$$

5.2 Tunable Single-Band Impedance Matching Network for Antennas

The tunability of C_1 is:

$$\tau_{C_1} \equiv \frac{C_{1max} - C_{1min}}{C_{1max}}. \tag{5.58}$$

As an example, for a T-network designed with symmetric impedance boundary, i.e. $R_L = 1/R_{La}$, the impedance matching range covered with varied tunability is shown in Fig. 5.44. Here, the initial Q_0 is set to 5 on the higher impedance boundary. The maximal varactor tunability varies from 20 % to 60 % accordingly.

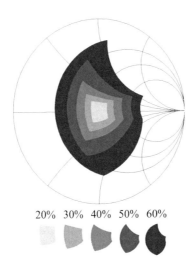

Fig. 5.44 Conjugate matching range at 2 GHz, and its dependence on varactor tunability

Choice of Network Q Factor

In a lossless T-network, when the network Q factor is high, there is a stronger change in transducer power gain (TPG) G_T in response to varactor tuning. Fig. 5.45 illustrates the dependence of G_T on the capacitance, where the Q_0 is set to 10 and 30. The swept capacitance is normalized to the solutions in Eqn. 5.49. When a network is tuned to a load of lower resistance, the corresponding network Q factor rises. It is clear that with higher network Q factor, the same tuning of varactor results in a larger change of G_T. It requires a more precise control of capacitance during tuning in order to seek the maximal gain, which imposes the fundamental constrains on the efficient control in section 5.2.5.

Furthermore, a higher network Q factor leads to stronger resonance in the network. There is more pronounced impedance change at the output port, even with the same varactor tunability. It leads to a larger conjugate impedance matching range of the TMN. As summarized in Fig. 5.46, a monotonous dependence of the matching range, i.e. the maximal attainable load resistance and reactance, on the network Q factor is observed. With a constant varactor tunability, e.g. 50 % here, at an increased Q_0 the TMN can match a larger reactance range, while the resistance range remains

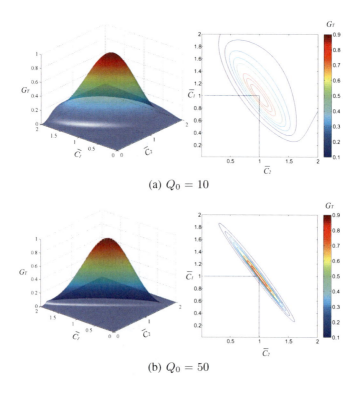

(a) $Q_0 = 10$

(b) $Q_0 = 50$

Fig. 5.45 Dependence of transducer power gain on varactors' capacitance: (a) $Q_0 = 10$, (b) $Q_0 = 50$. All the capacitances are normalized to their designs for a $150.5\,\Omega$ load.

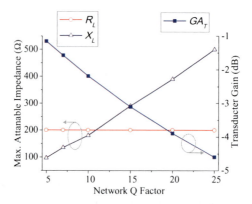

Fig. 5.46 The dependence of the matching range and the transducer power gain on the network Q factor, where a Q factor of 60 is assumed for varactors, and 65 for the inductor. The tunability of both varactors are set to 50 %.

constant. However, with high Q_0 the lossy components lead to higher insertion loss, which eventually reduces the transducer gain.

It is to conclude that the network Q factor is a determining factor of TMN' performance, including the choice of capacitance and inductance, as well as impedance matching range and sensitivity on varactor tuning. Therefore, the network design shall start at choosing a Q factor with the consideration of all above aspects.

5.2.2 Optimal Design with Lossy Components

When applying BST varactors and lumped inductors in the topologies in Fig. 5.39, which are no longer lossless, their loss is modeled as equivalent series resistance, namely R_{C_1}, R_{C_2} and R_{ind} in Fig. 5.47.

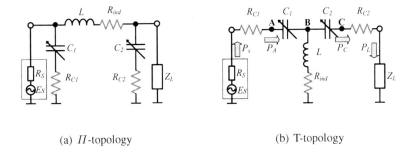

(a) Π-topology (b) T-topology

Fig. 5.47 Representative topologies of single-band tunable matching networks with lossy components

Such parasitic resistance invalidates the simultaneity of minimal reflection and maximal power transmission, which is the prerequisite of the above analysis. The optimal operation point has to be reconsidered, at the purpose of maximizing the transmission efficiency. In Fig. 5.47(b) P_S denotes the power available at the source, P_A the power available at point A, P_C the output power at the point C and P_L the power delivered to the load. By Thévenin's theorem [49], they are related as following:

$$\frac{P_A}{P_S} = \frac{R_S}{R_S + R_{C_1}}$$
$$\frac{P_L}{P_C} = \frac{R_L}{R_L + R_{C_2}}. \qquad (5.59)$$

It shows that, the overall transducer gain G_T is reduced compared to the gain G_{AC} by a linear factor, which is determined by the impedances and parasitic losses:

$$G_T \equiv \frac{P_L}{P_S} \text{ and } G_{AC} \equiv \frac{P_C}{P_A}$$

$$G_T = G_{AC} \frac{R_S R_L}{(R_S + R_{C_1})(R_L + R_{C_2})}. \quad (5.60)$$

For BST varactors working in paraelectric phase, the series resistance includes mainly the ohmic loss of electrodes and the dielectric loss in the BST film. Both can be treated as voltage independent. In this case, the condition of the maximal overall transducer gain is equal to the condition of maximal G_{AC}. When omitting the inductor's ohmic loss R_{ind}, the power gain G_{AC} is exactly the TPG of the lossless network between point A and C, while terminated with increased source and load impedance. By substituting G_S and G_L in Eqn. 5.49 and 5.50 with the following G'_S and G'_L, the operation points of C'_1, C'_2 and L' can be recalculated:

$$G'_S = \frac{1}{R_S + R_{C_1}}$$

$$G'_L = \frac{1}{R_L + R_{C_2}}. \quad (5.61)$$

Furthermore, at the angular frequency ω_m, the serial parasitic ohmic loss R_{ind} of the inductor can be transformed to conductance G_{ind} as illustrated in Fig. 5.48(a), which increases the conductance at point B and reduces the voltage at points B and C. Accordingly as shown in Fig. 5.48(b), the equivalent conductance G'_{BL} of modified capacitance C'_2 and resistance $1/G'_L$ shall be reduced to maintain the voltage at point B.

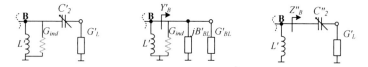

(a) Transformation of the inductor's loss.

(b) The loss increases the conductance at the right of point B.

(c) The parasitic resistance is absorbed by increasing load conductance.

Fig. 5.48 Compensation of inductor's parasitic loss, which increases the admittance load at point B

The compensation results in the circuitry in Fig. 5.48(c). The relations are given as:

$$\frac{1}{Z''_B} = Y'_b$$

$$\frac{1}{1/G''_L + 1/j\omega_m C''_2} = G_{ind} + G'_{BL} + jB'_{BL}. \quad (5.62)$$

5.2 Tunable Single-Band Impedance Matching Network for Antennas

The modifications result in a new design condition of the load, namely G_L'' instead of G_L':

$$G_L'' = G_{BL}' + G_{ind} + \frac{{G_{BL}'}^2 {G_L'}^2}{\omega_m^2 {C_2'}^2 (G_{BL}' + G_{ind})}, \quad (5.63)$$

where

$$G_{BL}' = \frac{\omega_m^2 {C_2'}^2 G_L'}{{G_L'}^2 + \omega_m^2 {C_2'}^2}$$

$$G_{ind} = \frac{R_{ind}}{R_{ind}^2 + \omega_m^2 L^2}. \quad (5.64)$$

By applying the G_L'' and G_S' in Eqn. 5.49 and 5.50, through an iteration the component values are optimized for the maximal TPG.

As an example, the Q factor of all varactors and inductor is assumed as 60 and the network Q factor $Q_0 = 4$ at 2 GHz. In order to match 150.5 Ω load, the optimal capacitances can differ up to 7 % from the lossless design. Consequently the TPG is improved by 0.11 dB.

Table 5.9 Exemplary comparison between lossless and modified designs of T topology matching network

	Lossless design	Modified design	Relative difference
C_1	0.3091 pF	0.3139 pF	1.6 %
C_2	0.1854 pF	0.1985 pF	7 %

Meanwhile, the impedances at the left and right sides of the point A are matched. When the input impedance at point A is denoted as $Z_{A_{in}}$, there is:

$$Z_{A_{in}} = R_S + R_{C_1}. \quad (5.65)$$

The reflection at the source is given by:

$$\Gamma_{in} = \frac{Z_{A_{in}} + R_{C_1} - R_S}{Z_{A_{in}} + R_{C_1} + R_S} = \frac{R_{C_1}}{R_{C_1} + R_S}. \quad (5.66)$$

It suggests a slight mismatch. In the above mentioned example, the mismatched input reflection is not higher than -18 dB, which is acceptable for mobile frontends. Hence, the benefits of using TPG in optimizing the lossy TMN have been clearly shown.

On the contrary, with the criterion of minimal input reflection the compensation in Eqn. 5.61 is modified as following, which substitutes G_S and G_L in Eqn. 5.49 and 5.50:

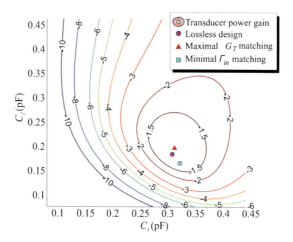

Fig. 5.49 Shift of the operation points with different matching criteria of the maximal transducer power gain and the minimal input reflection. The load resistance is 150.5 Ω.

$$G'_S = \frac{1}{R_S - R_{C_1}}$$
$$G'_L = \frac{1}{R_L - R_{C_2}}. \quad (5.67)$$

Then the network operates at different points. One shall notice that, sine no measure of the internal power loss is taken there shall be more than one operation point satisfying the reflection criterion. As compared in Fig. 5.49, in lossy networks the operation points depend on the design criterion. They impose constrains on the achievable TPG. Such influence is to be verified in section 5.2.5.

5.2.3 Antenna Bandwidth Enhancement

The above mentioned analyses restrict to single frequency and variable load impedance. However, as illustrated in Fig. 5.38 the antenna impedance is typically frequency-dependent. TMNs are expected to stabilize such variable impedance across a bandwidth. The following takes frequency dependent load into account, and illustrates TMN's potential for bandwidth enhancement of unmatched antennas.

At a single frequency f_0, the analytical solution of the lossless capacitance and inductance of the T-network is rewritten. As in Eqn. 5.49 and 5.50, it is reasonable to assume at first only resistance R_S and R_{Lf_0} of source and load at the frequency f_0, respectively. For chosen network quality factor Q_{f_0} at angular frequency ω_{f_0}, the solutions of C_{1f_0}, C_{2f_0} and L of the T-network are given by:

5.2 Tunable Single-Band Impedance Matching Network for Antennas

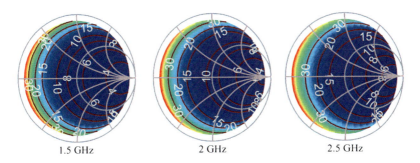

Fig. 5.50 Frequency dependence of network Q factor

$$\omega_{f_0} C_{1f_0} = \frac{(1 - K_{f_0})}{2Q_{f_0} - \sqrt{4Q_{f_0}^2 K_{f_0} - (1 - K_{f_0})^2}} \frac{1}{R_S}$$

$$\omega_{f_0} C_{2f_0} = \frac{(K_{f_0} - 1)}{2Q_{f_0} - \sqrt{4Q_{f_0}^2/K_{f_0} - \left(1 - 1/K_{f_0}^2\right)^2}} \frac{1}{R_S}$$

$$\omega_{f_0} L = \frac{2Q_{f_0}(K_{f_0} + 1) - 2\sqrt{4Q_{f_0}^2 K_{f_0} - (K_{f_0} - 1)^2}}{(K_{f_0} - 1)^2} R_S, \quad (5.68)$$

where $K_{f_0} = R_S/R_{Lf_0}$.

When static inductor is used, once the Q_{f_0} and L are designed for load R_{Lf_0}, the network quality factor Q_{f_i} for an arbitrary load R_{Lf_i} at frequency f_i can not be chosen but is determined by:

$$Q_{f_i} = \frac{(1/K_{f_i} + 1) + \sqrt{(1/K_{f_i} + 1)^2 - 4R_{Lf_i}/(L\omega_{f_i})^2 - (1/K_{f_i} - 1)^2}}{2R_{Lf_i}/(L\omega_{f_i})}, \quad (5.69)$$

where $K_{f_i} = R_S/R_{Lf_i}$, and R_{Lf_i} is the load resistance.

As an example in Fig. 5.50, the network Q factor is initially assigned as 6 to match a 150.5 Ω load at 2 GHz. For varied load at either the same or different frequency, the Q factor is determined due to the static inductor. The Q factor at 1.5 GHz and 2.5 GHz for the same load is 4.5 and 7.8, respectively. Accordingly the capacitors required to match R_{Lf_i} are obtained by:

$$\omega_{f_i} C_{1f_i} = \frac{(1 - K_{f_i})}{2Q_{f_i} - \sqrt{4Q_{f_i}^2 K_{f_i} - (1 - K_{f_i})^2}} \frac{1}{R_S}$$

$$\omega_{f_i} C_{2f_i} = \frac{(K_{f_i} - 1)}{2Q_{f_i} - \sqrt{4Q_{f_i}^2/K_{f_i} - \left(1 - 1/K_{f_i}^2\right)^2}} \frac{1}{R_S}, \quad (5.70)$$

which use the drifted Q factor Q_{f_i}.

The frequency dependent load, e.g. electrically small antennas, can be modeled as a resonator. In Fig. 5.51, its input reflection is simplified to a constant level namely Γ across the frequency range from f_0 to f_4. The reflections are $\Gamma, -j\Gamma, -\Gamma, j\Gamma$ and Γ for the frequency f_0, f_1, f_2, f_3 and f_4 at equal intervals. The network is initially designed at f_0 by Eqn. 5.68. Then by Eqn. 5.70, the capacitance to match the load resistance R_{Lf_i} is obtained, while an absorption method is used to compensate the reactance jX_{Lf_i}:

$$\frac{1}{C'_{f_i}} = \frac{1}{C_{f_i}} + \omega_{Lf_i} X_{Lf_i}, \qquad (5.71)$$

where $jX_{Lf_i} = Z_{Lf_i} - R_{Lf_i}$, and $i = 0\ldots 4$.

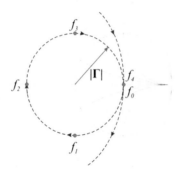

Fig. 5.51 Simplified model of frequency-dependent load of a single-band antenna

The necessary tuning of the capacitance is achieved by using varactors, the tunability τ_{max} of which is defined as a relative capacitance change:

$$\tau_{C1max} = \frac{max_{i=0\ldots 4} C'_{1f_i} - min_{i=0\ldots 4} C'_{1f_i}}{max_{i=0\ldots 4} C'_{1f_i}}$$

$$\tau_{C2max} = \frac{max_{i=0\ldots 4} C'_{2f_i} - min_{i=0\ldots 4} C'_{2f_i}}{max_{i=0\ldots 4} C'_{2f_i}}. \qquad (5.72)$$

For given fractional bandwidth BW_f, as defined in Eqn. 5.73, and different reflection level Γ, the required tunability τ_{max} of each varactor is shown in Fig. 5.52. The intercept points at negligible reflection level e.g. $\Gamma = -50$ dB illustrate the necessary tuning to compensate the components' reactance across the target bandwidth. When either the bandwidth or antenna's reflection level increases, the required tunability increases.

$$BW_f = \frac{f_4 - f_0}{\sqrt{f_0 f_4}} \qquad (5.73)$$

5.2 Tunable Single-Band Impedance Matching Network for Antennas

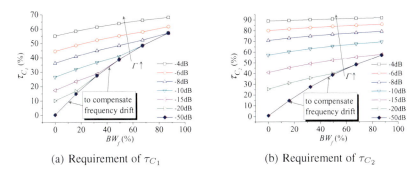

(a) Requirement of τ_{C_1}

(b) Requirement of τ_{C_2}

Fig. 5.52 Required tunability of the two varactors in T-network. Lines are grouped for different reflection levels. The band is centered at 2 GHz.

A ferroelectric TMN has been implemented to match a poorly performing monopole antenna, where $C_1 = 0.57$ pF, $C_2 = 0.5$ pF and $L = 8.1$ nH. As in Fig. 5.53, the kernel TMN encompasses a piece of ceramic module with two IDC varactors and an external discrete inductor. The ceramic module has input and output contact pads as well as a central pad to connect inductor. All pads are connected through silver epoxy glue. It is mounted on the ground of an unmatched monopole antenna, with its pads facing upward. Discrete decoupling capacitors and resistors are soldered at the ports of the TMN as bias-tees. The whole circuit has six discrete components.

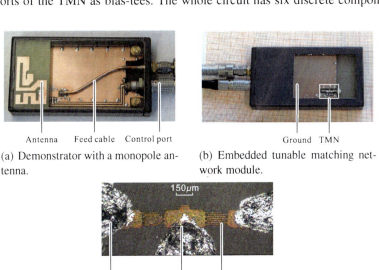

(a) Demonstrator with a monopole antenna.

(b) Embedded tunable matching network module.

(c) Ceramic module with a tunable interdigital capacitor.

Fig. 5.53 Demonstrator of antenna bandwidth enhancement with a tunable matching network on BST thick-film

A 50 Ω coaxial cable feeds the antenna through the TMN. A pair of bias voltage is generated by an external source and delivered through an additional control port. The module including peripherals has a footprint of 8.1 mm × 8.5 mm. The antenna is measured to be 35 mm × 20 mm and 8 mm extruding above a ground plane of 40 mm × 75 mm.

The original antenna exhibits a reflection between -3.5 dB and -6 dB from 1.5 GHz to 1.8 GHz. As shown in Fig. 5.54, the varactors are tuned by 0 % to 41 % independently. The best matched frequency shifts accordingly. A general improvement of reflection is achieved from 1.35 GHz to 1.98 GHz, especially the reflection is below -15 dB over the range from 1.48 GHz to 1.81 GHz. The instantaneous bandwidth at -15 dB is from 57 MHz to 140 MHz.

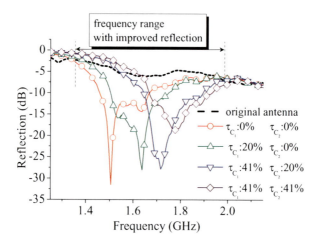

Fig. 5.54 Measurement of bandwidth enhancement of the demonstrator in Fig. 5.53

5.2.4 System in Package Realization

The prototype consists of multiple discrete components. An integrated module within a single package is desired. Apart from using bonding wires or conductive epoxy glue, the parasitics in the module can be mitigated through solder bumps and flip chip technology. Here a discrete off-chip inductor is implemented instead of integrated one. Firstly, high Q factor is indispensable for the TMN. Secondly, by isolating the varactor module and inductor, one gains a flexibility in prototyping and fine tuning, while the knowledge of which can be transferred to full SIP designs. Lastly, the concept of the packaged module is compatible with LTCC carriers, which can be extended to embed high Q factor inductor in the carrier [98]. In the following, several technical challenges towards a packaged module are identified and coped with. The efforts lead to a demonstrative prototype in Fig. 5.55. The module consists of the integrated varactor module and an inductor. Both are flip-chip mounted on a multilayer carrier. Its functionality is finally proven through verification.

5.2 Tunable Single-Band Impedance Matching Network for Antennas

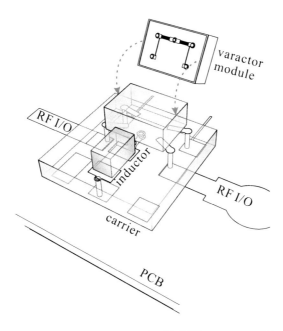

Fig. 5.55 Integrated tunable matching network with BST varactor module and lumped inductor. It is assembled into a multilayer package.

Integrated Resistive Bias Network

The kernel design is evolved from that in section 5.2.3. The off-chip bias tees are integrated into the module by two means. First, the low-pass resistive branch is realized through ITO strip lines, which are introduced in section 3.1.2. Second, the function of a high-pass DC-block can be achieved in an anti-serial varactor pair where the bias is fed at the middle of the two varactors while the outer ends of the varactors are DC-grounded through resistors. The topology is illustrated in Fig. 5.56.

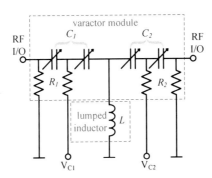

Fig. 5.56 Integration of bias-tee into tunable matching network module

Four strip lines of 400 μm length and 10 μm width are used for the R_1 and R_2. In [83], varactor's series resistive parasitics deteriorates its Q factor and compromise the insertion loss. In the case of parallel parasitics here, the influence is the reduction of matching range. In a circuitry simulation, a 40 % varactor tunability is assumed. When increasing the resistivity of the strips from 50 Ω/sq, the achievable conjugate matching range at 2 GHz increases and approaches to the maximum at 1000 Ω/sq which is close to the resistivity of 20 nm-thick ITO. A comparison is depicted in Fig. 5.57. It comes to the conclusion that ITO can be conveniently used for the biasing purpose, while the tolerance in oxidization state may impose a minor influence on the module performance.

With the processing technologies in section 3.1, a varactor module is fabricated with integrated ITO bias network. As in Fig. 5.58, it has 6 contact pads, including two I/O ports, one pad to off-chip inductor and two pads for biasing. The bottom central pads can feed voltages. The off-chip inductor grounds through a DC-block

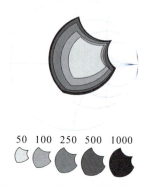

Fig. 5.57 Influence of resistive bias network on the matching range. The corresponding sheet resistance is listed with units of Ω/sq.

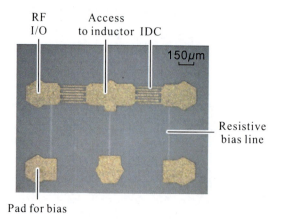

Fig. 5.58 Varactor pair with resistive bias network in a ceramic module

5.2 Tunable Single-Band Impedance Matching Network for Antennas

capacitor, which is only for debug purpose. The anti-serial varactor topology and the resistive grounding are not implemented, in order to reduce complexity but without loss of generality.

Multilayer Carrier

Multilayer LTCC is one of the popular SIP technologies. In addition to the interconnects, high Q factor embedded inductors have also been demonstrated [98]. Considering the cost and lead time of LTCC prototyping, a multilayer laminate carrier is implemented for evaluation. Geometries of the proposed carrier can be found in Fig. 5.59(b). Compared to the epoxy bonding in the demonstrator in section 5.2.3, the vias and interconnects here are considerable shorter, which helps to reduce the inductive and resistive parasitics. The varactor module is mounted on the surface of the carrier through flip-chip technologies as in Fig. 5.59(a). The air gap at the bumps can be sealed with dielectric underfill. Considering the air gap is relatively thin and the underfill has a permittivity lower than 3.5, its effect is therefore negligible.

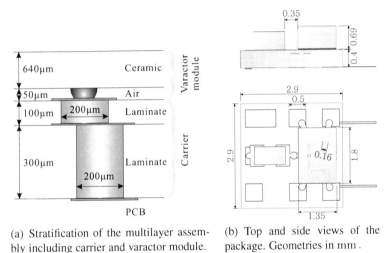

(a) Stratification of the multilayer assembly including carrier and varactor module.

(b) Top and side views of the package. Geometries in mm.

Fig. 5.59 Multilayer carrier with embedded connections

The whole module is compatible with ball grid array (BGA) package. With the dimensions, its parasitics mainly come from three sources. First, the inductance of the vias and strip lines in serial connection with the varactor, the off-chip inductor and the bias network. Second, the shunted capacitive coupling from the I/O ports to ground. Third, the mutual coupling between input and output ports. All the parasitics are modeled in the equivalent circuit in Fig.5.60. The laminate layers have a relative permittivity of 4. By taking the finite element method simulation tool Sonnet, the components' values are approximated using least mean squares (LMS) method across the frequency range from 1.9 GHz to 2.1 GHz, as listed in Table 5.10. The metallic loss of the strips is not considered here.

Table 5.10 Parasitics in the equivalent circuit in Fig. 5.60

L_{P1}	L_{P2}	C_{P1}	C_{P2}	C_{PM}	L_1	L_2	L_L
0.5 nH	0.5 nH	2.5 fF	2.5 fF	2 fF	0.8 nH	0.8 nH	0.6 nH

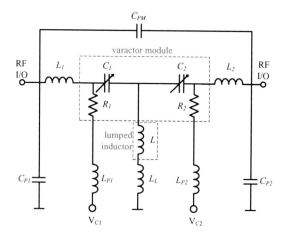

Fig. 5.60 Equivalent circuit of the integrated tunable matching network module

The kernel TMN design is slightly tuned accordingly. The expected conjugate matching range across the frequency range is illustrated in Fig. 5.61 with a 40 % varactor tunability.

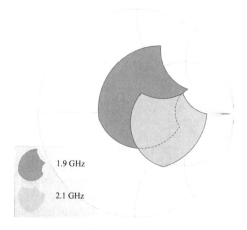

Fig. 5.61 Simulated conjugate matching ranges at 1.9 GHz and 2.1 GHz, respectively

5.2 Tunable Single-Band Impedance Matching Network for Antennas

Flip-Chip Assembly

The under bump material (UBM) is chosen to be Ti and Au according to the optimization in section 3.1.2. Gold ball bumps are afterwards bonded on the contact pads as in Fig. 5.62(a). The balls are reflowed under 235 °C. Moderate adhesion of the metalization guarantees an affordable yield ratio. The bonded varactor modules are then flip-chip mounted onto the carrier. The optimized pressure is 41 cN. The yield ratio here is mainly determined by the limited adhesion between the UBM and the BST layer during flip-chip process.

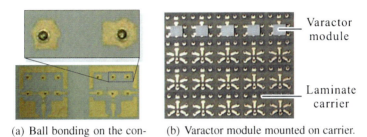

(a) Ball bonding on the contact pads.

(b) Varactor module mounted on carrier.

Fig. 5.62 Ball bonding and flip-chip assembly of the BST thick-film varactor module

Prototype Verification

A 5.6 nH high Q inductor from Murata is soldered on the pre-assembly, which is then populated on a PCB testbed. As in Fig. 5.63, the RF input and output ports as well as the DC bias ports are connected to edge-mounted subminiature version a (SMA) connectors through 50 Ω microstrip lines. The electric lengths of the strips from the RF I/Os are de-embedded, therefore, the reference plane is at the soldering pads beneath the TMN module.

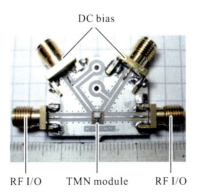

Fig. 5.63 Packaged tunable matching network module in testbed

From a Keithley 2612A source meter, two independent channels of DC bias voltage from 0 V to 100 V are applied to the varactors. The two port scattering matrix of the module is measured with 50 Ω reference on Anritsu 37397C VNA. The input reflection and transmission coefficients are measured during the tuning. When both varactors are biased from 0 V to 100 V, the TMN exhibits a tunable pass band across the target frequency range from 1.8 GHz to 2.1 GHz as shown in Fig. 5.64. With a Q factor of 60 of the varactors and 64 of the inductor, the insertion loss with 50 Ω source and load is found between 0.86 dB and 0.98 dB.

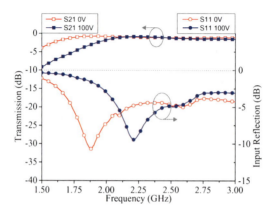

Fig. 5.64 Measured scattering coefficients with 0 V and 100 V bias voltage on both varactors

When varying the bias voltages, the output impedance of the TMN is extracted from a post-simulation in Agilent Advanced Design System as illustrated in Fig. 5.65. Its conjugate complex matching range is found. Furthermore by sweeping the source impedance along the circle of constant reflection level Γ=-10 dB, the corresponding matching range can be found at the load. The actual tunability of the varactors in the demonstrator module is found to be 26 %, which leads to considerably reduced ranges compared to the simulation in Fig. 5.61. The frequency dependence of the matching ranges from 1.8 GHz to 2.1 GHz is illustrated. The aimed functionality of the packaged module is therefore clearly proven. However, the offset of the matching range towards inductive domain suggests residual capacitive parasitics in the assembly, which shall be compensated by further reducing the varactors' capacitance.

The integrated TMN is used to match the antenna measured in Fig. 5.38. Under various conditions the TMN can always stabilize the impedance at its input port. Such stabilization across multiple frequency bands is demonstrated in Fig. 5.66.

5.2.5 Efficiency of Adaptive Control

In order to maximize the networks' performance under changing conditions, an adaptive control of the varactors is indispensable. It has to be simple and convergent to an optimum within reasonable seeking period. The following focuses on

5.2 Tunable Single-Band Impedance Matching Network for Antennas

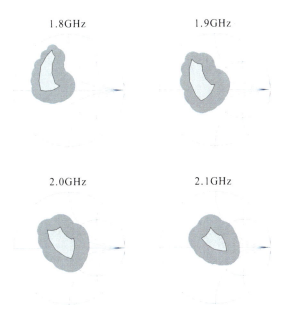

Fig. 5.65 Measurement of the matching ranges: conjugate matching range marked by light gray with solid boundary; and -10 dB matching range in darker gray

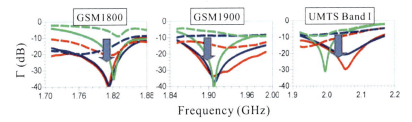

Fig. 5.66 Measurements of reflection reduction across multiple frequency bands in various antenna operation states. All the dash lines denote the reflection without a tunable matching network. The solid lines indicate the improved reflection with the tunable matching network.

the control of lossy tunable matching networks with two different matching criteria: minimal reflection at the input port of the matching network or maximal output power delivered to the load. The constraints from the network quality factor Q_0 and resolution of the digital to analog converter (DAC) for the generation of tuning voltages are illustrated. The control's efficiency is finally demonstrated on the prototype realized in section 5.2.4.

Fig. 5.67 compares two subsystem architectures for the matching criterion of minimal reflection and maximal TPG respectively. They consist of the TMN itself, detector of either reflection coefficient or TPG, signal processing part, which can be an embedded microcomputer or integrated in the baseband processor, and converters between digital and analogue domains.

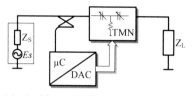 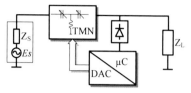

(a) Architecture for minimal reflection matching.

(b) Architecture for maximal transducer power gain matching.

Fig. 5.67 Block diagrams of the subsystems for different matching criteria

Following the Eqn. 5.60, for linear load the TPG G_T is linearly related to the voltage transformation ratio $H(j\omega) = V_L/E_S$ across the TMN, where V_L is the voltage across the load and E_S denote the source voltage [112]:

$$G_T = 4\frac{Z_S}{Z_L}|H(j\omega)|^2 . \qquad (5.74)$$

It allows the use of a simple RF voltage detector at TMN's output port rather than a bulky and lossy directional coupler at the input port. The diode detector rectifies the RF voltage. A conjugate gradient algorithm is implemented in a micro controller, which uses a steepest descent approach to find the extreme value of the given cost function [23] i.e. the maximal RF voltage at load. After an initialization step it seeks the maximal V_L by varying the bias voltages. The digital output of the controller is converted by DAC to the analog voltage and then amplified by a DC-DC converter to tune the varactors. A singular maximum of G_T guarantees a convergent control. As shown in Fig. 5.68(a), for the low-pass Π-matching network, there is another local maximum of $|H(j\omega)|$ at $\omega C_1 = \omega C_2 = 0$, while for the high-pass T-matching network, the other maximum appears at infinite. In this case, gradient method shows a higher robustness in converging to the global maximum in T-network rather than in Π-network.

In section 5.2.1, the determining influence of network Q factor namely Q_0 on the tuning sensitivity of G_T is discussed. In other words, with a higher Q_0 the G_T is more subject to the varactors' tuning. The constraint leads to different convergence speed as illustrated in Fig. 5.69(a). By taking again the example in section 5.2.1, the TPG at 2 GHz is denoted in contours in relation to the varactors' capacitance, where the global maximum is marked in upward-pointing triangle. With a low Q_0 of 4 at the 150.5 Ω load, the G_T changes smoothly, which results in a quick convergence within 2 iterations from the center of varactors' tuning range. It is shown in Eqn. 5.51 that when the inductor is fixed, the Q_0 turns to depend on the load impedance. Once the Q_0 and L are designed for a load Z_L, the Q_{0a} for another load Z_{La} is subject to this determinative dependence. In T-network, for Z_{La} with a resistance lower than that of Z_L, the Q_{0a} is higher than Q_0. When the above mentioned network is tuned to match a 16.6 Ω load, Q_{0a} is 12.8. In Fig. 5.69(b), the gradient method has to repeatedly adjust the estimation of the steepest descent direction and refine the diameter of the trust region, which cost 4 iterations. It is clear that the same TMN may exhibit different convergence speed on varied load.

5.2 Tunable Single-Band Impedance Matching Network for Antennas   113

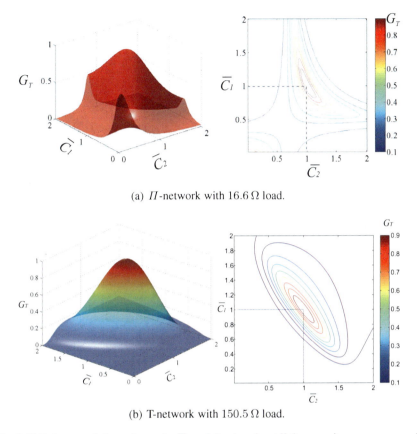

(a) Π-network with 16.6 Ω load.

(b) T-network with 150.5 Ω load.

Fig. 5.68 Existence of the optimum in Π- and T-networks. All the capacitances are normalized to their designs with a same network Q factor $Q_0 = 10$.

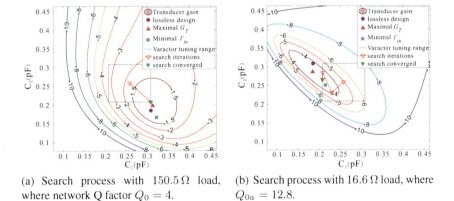

(a) Search process with 150.5 Ω load, where network Q factor $Q_0 = 4$.

(b) Search process with 16.6 Ω load, where $Q_{0a} = 12.8$.

Fig. 5.69 Simulated convergent search of maximal transducer power gain. The simulation is done at 2 GHz on T-network. The tunability of varactors is set to be 30 %.

The application of DAC introduces further constraints on both the convergence speed and the achievable TPG. As shown in Fig. 5.70(a) when using a 4 bit DAC to generate a single polarization bias voltage, the varactor hops its capacitance between discrete tuning states. With low DAC resolution the gradient method may not be able to reduce the trust region's diameter, resulting in a poor approximation to the global optimum. The adaptive control converges to a reduced TPG, which is compared to the maximal attainable gain.

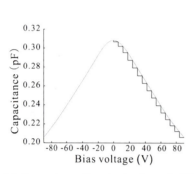
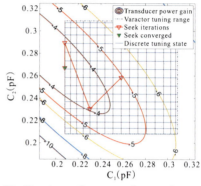

(a) Exemplary discrete tuning of BST thick-film varactor, which is denoted by the solid step line. The varactor is only positively biased.

(b) Hopping of varactors' capacitances when seeking the maximal transducer power gain.

Fig. 5.70 Discrete tuning states limited by the resolution of digital to analog converter

Meanwhile, with a higher network Q factor the TPG is more sensitive to the varactors' tuning. As a result the algorithm converges away from the optimum and hence reduces the gain. In Fig. 5.71, with various network Q factor the attainable G_T always improves with an increased DAC resolution. It also shows that a TMN with higher Q_0 require a higher DAC resolution. For example, with an affordable

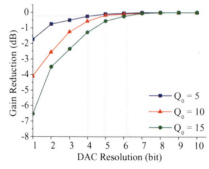

Fig. 5.71 Dependence of transducer gain reduction on DA-Converter resolution and network Q factor. A lossless tunable matching network is assumed. In this case, the transducer power gain reduces only due to various convergence of discrete tuning.

5.2 Tunable Single-Band Impedance Matching Network for Antennas

-0.25 dB gain reduction, a TMN with $Q_0 = 5$ requires 4 bits DAC, while 5 bits for $Q_0 = 10$ and 6 bits for $Q_0 = 15$.

As a demonstration, the control is carried out on the prototype in section 5.2.4 at 1.89 GHz. In order to set the load impedance, an automatic passive load-pull tuner from Focus Microwaves Inc. is used. An Anritsu VNA measures the two port scattering matrix. The corresponding voltage transformation ratio and TPG is calculated in a Matlab® code, where the conjugate gradient method is running too. An external dual channel voltage source drives the TMN with the bias voltage assigned by the control algorithm. The setup is shown in Fig. 5.72. The setup is able to evaluate both two optimization criteria and the influence of DAC resolution.

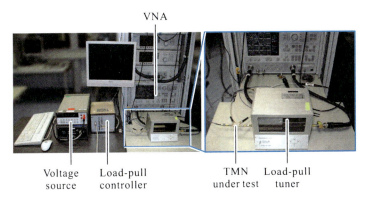

Fig. 5.72 Setup for load-pull verification of a tunable matching network

As shown in Fig. 5.73(a), the control seeks the maximal transducer gain. With an arbitrary load impedance of e.g. $28+j40\,\Omega$ and a 9-bit DAC resolution, a TPG of -1.01 dB and a reflection of -18.6 dB are achieved within 5 iterations. In other words, besides the input impedance matching, there is 0.4 dB improvement of the power delivered to the load compared to a direct connection without the TMN. In Fig. 5.73(b) the minimal reflection criterion is applied. The load and the DAC are

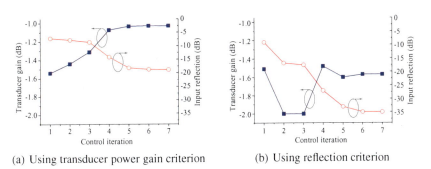

(a) Using transducer power gain criterion (b) Using reflection criterion

Fig. 5.73 Measurement of the adaptive control process of the tunable matching network. The performances of using different control criterion are compared.

the same as above. In 6 iterations, a low reflection of -35 dB is achieved, with however a compromised TPG of -1.55 dB. These measurements agree well with those analyses in section 5.2.2. It can be seen that, the maximal TPG criterion exhibits better balance between the reflection and the gain.

The advantage of using TPG to optimize TMN with lossy BST varactors is demonstrated. Together with the design guide in section 5.2.1, a comprehensive knowledge on the design, realization and utilization of single band TMN specifically T-topology has been presented. In the following, innovative TMNs for multi-band applications are addressed.

5.3 Tunable Multiband Matching Network

All the above tunable matching networks operate in single frequency bands. They can provide an improved impedance matching and power transmission within a certain bandwidth. When they are forced to work beyond their bandwidth, great part of the tunability will be utilized to compensate the reactance drift, as described in section 5.2.3. However, the wireless services may have discrete spectrum allocations, which are typically distant from each other. Therefore, TMN are desired to operate simultaneously in multiple bands instead of a single one. Meanwhile, the design methods above and in the reference works provides only guidelines in the circuit level for specific topologies. Both the optimum and performance are therefore not guaranteed. In this part, polynomial level design methods are to be discussed and modified for multiband matching problems. The compensation of components' loss is further realized at the polynomial synthesis stage.

5.3.1 Polynomial Synthesis Method for Impedance Matching

The method discussed here follows the real frequency technique (RFT) which was proposed in [13, 14] and then extended in [17] for the broadband double matching problems. The essence of the RFT is to maximize the TPG. Here, the RFT is modified into a multiband design method, named as modified real frequency technique or MRFT. Like in the original RFT, there is no need to assume either the topology or some initial component values for the network in the MRFT.

The TPG is a function of the Thévenin's impedance seen from the load side, defined as Z_q in Fig. 5.74. For the given source and load impedances, Z_S and Z_L, the TPG is rewritten [30]:

$$GA_T(\omega) = \frac{\left(1 - |g_{22}|^2\right)\left(1 - |s_{in}|^2\right)}{|1 - g_{22}s_{in}|^2}, \qquad (5.75)$$

5.3 Tunable Multiband Matching Network

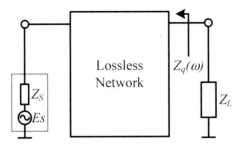

Fig. 5.74 Doubly terminated lossless network

where

$$s_{in} = e^{2j\phi_F(\omega)} \left(\frac{Z_L - Z_q(-j\omega)}{Z_L + Z_q} \right)$$

$$g_{22}(j\omega) = \frac{Z_S - 1}{Z_S + 1} \cdot$$

The $\phi_F(\omega)$ is the phase function. It is related with the network transmission zeros [14]. The TPG depends on the source impedance Z_S, the load impedance Z_L and the impedance looking from the load side to the network, Z_q. The source and load impedances are given as design targets. Thus, Thévenin's impedance, $Z_q(\omega) = R_q(\omega) + jX_q(\omega)$ can be described in terms of a number of unknown real parameters. These unknowns are determined by optimizing the TPG.

At the frequencies where the TPG is optimized, there is:

$$|g_{22}s_{in}| \ll 1 . \tag{5.76}$$

For such perfect matching, the following relation is tenable [17]:

$$[R_q(\omega) + R_L(\omega)]^2 \gg [X_Q(\omega) + X_L(\omega)]^2 . \tag{5.77}$$

Therefore, the TPG can be simplified:

$$GA_T(\omega) = \frac{4R_L(\omega)R_q(\omega)\left(1 - |g_{22}|^2\right)}{[R_q(\omega) + R_L(\omega)]^2} . \tag{5.78}$$

All the terms in the $G_0(\omega)$ are known, except $R_q(\omega)$. In the original RFT it is calculated as a frequency dependent impedance across a continuous band by:

$$R_q(\omega) = R_L(\omega)\left[K(\omega) \mp \sqrt{K(\omega)^2 - 1}\right] , \tag{5.79}$$

where

$$K(\omega) = \frac{2\left(1 - |g_{22}|^2\right)}{G_0(\omega)}. \tag{5.80}$$

Here, a line segment representation of $R_q(\omega)$ is utilized. As in [12], straight-line segments are used to find $R_q(\omega)$ such that it is assumed to be composed of semi-infinite slopes with frequency break points at $0 < \omega_1 < \omega_2 < \ldots < \omega_n$. The break points $\omega_k, k = 1, \ldots, n$ are the frequencies where TPG is maximized. The points are chosen with more insight by qualitatively examining the real load across the frequency range. The real part of the Thévenin's impedance, namely $R_q(\omega)$, can be specified as a linear combination of the individual total resistive excursions, namely r, of each of the straight line segments:

$$R_q(\omega) = r_0 + \sum_{k=1}^{n} a_k(\omega) r_k = r_0 + a^T r. \tag{5.81}$$

The weight coefficients a_k and the segment resistance r_k are the unknowns.

In order to decrease reflections, matching must be done only in the bands of interest. Accordingly, it requires $R_q(\omega) = 0$ for $\omega > \omega_n$:

$$r_0 + \sum_{k=1}^{n} r_k = 0. \tag{5.82}$$

Thus, there are n unknowns. Actually it is possible to calculate r_0 directly by $r_0 = R_q(\omega = 0)$. Therefore, the number of unknowns is reduced by one. The a_k can be chosen as:

$$a_k(\omega) = \begin{cases} 1, & \omega \geq \omega_k \\ \dfrac{\omega - \omega_{k-1}}{\omega_k - \omega_{k-1}}, & \omega_{k-1} \leq \omega \leq \omega_k \\ 0, & \omega \leq \omega_{k-1} \end{cases}. \tag{5.83}$$

The impedance can be calculated from its real part by using Hilbert transform method. By combining Hilbert transform method and line segment representation, the reactance $X_q(\omega)$ can be specified by these unknown recursions, r_k.

$$X_q(\omega) = \sum_{k=1}^{n} b_k(\omega) r_k = b^T r, \tag{5.84}$$

where

$$b_k(\omega) = \frac{1}{(\omega_k - \omega_{k-1})\pi} \int_{\omega_{k-1}}^{\omega_k} \ln\left|\frac{y + \omega}{y - \omega}\right| dy. \tag{5.85}$$

In [15], through the line segment representation, the following relations are achieved. First, the gain at any frequency is weakly affected by the contributions of line segments away from the frequency of interest. Second, when the total reactance $X_t(\omega) = X_q(\omega) + X_L(\omega)$ is less than $0.3(R_q(\omega) + R_L(\omega))$, its effect is

5.3 Tunable Multiband Matching Network

insignificant since the impedance $Z_q + Z_L$ is approximately equal to the total resistance. Third, the arithmetic is very simple and the gain depends directly on the unknown r_k. Finally, the poles of the rational function approximating to $R_q(\omega)$ form the initial polynomial.

The second step of MRFT is to find an appropriate rational function $\bar{R}_q(\omega)$ for the calculated $R_q(\omega)$ which is the even part of $Z_q(\omega)$. Hence, $Z_q(\omega)$ has the same Hurwitz denominator and nominator as $R_q(\omega)$. In [17], the transmission zeros are set at infinity, so $R_q(\omega)$ has the simplest representation with its numerator always as one. This assumption simplifies the calculations, but since these zeros are the transmission zeros of the networks, it leads to the low-pass topologies. In this case, the following representation of $R_q(\omega)$ is [17]:

$$\bar{R}_q(\omega) = \frac{1}{D(j\omega)D(-j\omega)} = \frac{1}{\bar{D}(\omega^2)} = \frac{1}{A_0 + B_1\omega^2 + \ldots + B_n\omega^{2n}}. \quad (5.86)$$

The $D(j\omega)$ is a Hurwitz polynomial. These physical realization conditions add further limitations over the frequency range. $\bar{R}_q(\omega)$ can be approximated through a constrained curve fitting algorithm as:

$$\bar{D}(\omega^2) = 0.5\left[P_n^2(\omega) + P_n^2(-\omega)\right], \quad (5.87)$$

where

$$P_n(\omega) = 1 + x_1\omega + \ldots + x_n\omega^n. \quad (5.88)$$

Then $\bar{D}(\omega^2)$ is nonnegative and even. The relation between the coefficients x_i and A_0, B_i can be found as following:

$$\begin{aligned}
A_0 &= x_0^2 \\
B_1 &= x_1^2 + 2x_2 \\
&\vdots \\
B_k &= x_k^2 + 2\left(x_{2k} + \sum_{j=2}^{k} x_{j-1} x(2k-j+1)\right) \\
B_n &= x_n^2.
\end{aligned} \quad (5.89)$$

Finally, the corresponding $\bar{D}(\omega)$ is generated and this polynomial is the output of the second step.

However, in the case of multiband matching, it is desirable to increase the matching range across the band of interest while suppressing the out-band transmission. The new types of the circuitry units can be added to the network by adding different types of zeros to the polynomials. Therefore, the appropriate rational function $\bar{R}_q(\omega)$ is preferred to have not only the zeros at infinity but also finite zeros, which leads to bandpass topologies, or zeros at the origin which means high-pass topologies. Hence the following equation gives the $R_q(\omega)$ in the MRFT:

$$\bar{R}_q(\omega) = \frac{A_k \omega^{2k}}{\omega^{2n} + b_{n-1}\omega^{2(n-1)} + \ldots + b_1\omega^2 + b_0}, \tag{5.90}$$

where $k = n/2$ gives good results. The new transmission zeros also affect the phase function in Eqn. 5.76.

The initial polynomial for the real part of $Z_q(\omega)$ is found in the above two steps. The transducer power gain is then calculated by Eqn. 5.78. Furthermore, since the remaining part of the impedance polynomial can be determined from its real part with the Gewertz method [17], the assumption in Eqn. 5.76 turns to be unnecessary. Hence, a general representation of TPG can be:

$$GA_T(\omega) = \frac{4R_L(\omega)R_q(\omega)\left(1 - |g_{22}|^2\right)}{[R_q(\omega) + R_L(\omega)]^2 + [X_q(\omega) + X_L(\omega)]^2}. \tag{5.91}$$

In contrast to the original RFT, firstly the positive and real $\bar{R}_q(\omega)$ polynomial is found. The $R_q(\omega)$ values are calculated then. Depending on the given problem, the computational effort is increased compared to the RFT, whereas the numerical stability of the cost function is maintained between the steps. Finally, both the assumptions in Eqn. 5.76 and 5.77 are dropped and the general formula given in Eqn. 5.91 is taken in optimization. In all steps the Levenberg-Marquart algorithm is used for the optimization. As a result, the Thévenins' impedance polynomial that maximizes the TPG is obtained and the network is realized.

As an example, three dual-band networks following the MRFT are designed at 0.9 GHz and 2 GHz. With 50 Ω source and the same randomly chosen load impedance, several topologies are designable as shown in Fig. 5.75.

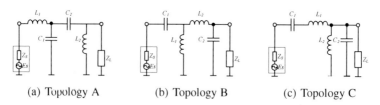

(a) Topology A   (b) Topology B   (c) Topology C

Fig. 5.75 Exemplary networks for dual-band matching following MRFT

A network following the original RFT with the same source and load conditions is also designed. Fig. 5.76 compares it with the three MRFT designs. It shows that the RFT network has a broadband characteristic over the entire frequency range. As a result, the maximal TPG obtained is -0.35 dB at 0.45 Hz. However, the MRFT enables the design of dual-band networks, and the maximal TPG is -0.132 dB at the desired frequencies.

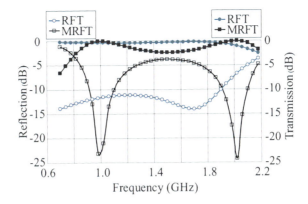

Fig. 5.76 Comparison of transmission and reflection of networks synthesized by the real frequency technique and the modified real frequency technique

5.3.2 Polynomial Optimization for Component Loss Compensation

In the following analysis, normalized scattering parameters of the lossless network are defined in Belevitch representation [9]:

$$\begin{bmatrix} s_{11} & s_{12} \\ s_{21} & s_{21} \end{bmatrix} = \frac{1}{g(s)} \begin{bmatrix} h(s) & \pm f(s) \\ \pm f(s) & -(-1)^k h(-s) \end{bmatrix}, \quad (5.92)$$

where k is an integer and equal to the order of the transmission zeros [17]. Since the network is lossless, the polynomials can be linked by the Feldtkeller equation [110]:

$$g(s)g^*(s) = f(s)f^*(s) + h(s)h^*(s), \quad (5.93)$$

where $*$ denotes conjugation of the polynomials. Here, the polynomials have real coefficients and the roots of the $g(s)$, namely Hurwitz denominator, are located in the left half of the complex frequency plane. These conditions must be satisfied to realize the networks from the polynomials. When two of the three polynomials are known, the unknown one is found using Eqn. 5.93. Afterwards, the Belevitch representation is formed. This representation is enough to analyze the networks and then find the optimal polynomials.

With the $g(s)$ and $h(s)$, the input driving-point impedance can be determined as:

$$Z(s) = \frac{1 + s_{11}}{1 - s_{11}} = \frac{g(s) + h(s)}{g(s) - h(s)}. \quad (5.94)$$

The network can be designed with the Darlington synthesis technique [16], using the $Z(s)$ polynomial.

When considering the component loss, the design networks can be further optimized in order to compensate the loss. In the following, the component loss can be estimated at polynomial level while the polynomials can be optimized for a lossy network. Additionally, since the improvement is achieved by the optimal polynomials, it is applicable for all realizable topologies of $Z(s)$ polynomial.

In the lossless MRFT networks the components are substituted for lossy components. Such a component has an equivalent series resistance, namely R. It is calculated for given component Q factor at the frequency f_0:

$$R_i f_0 = \frac{|X_i(f_0)|}{Q_i} . \qquad (5.95)$$

The index i refers to the individual capacitors and inductors. X is the reactance. For simplicity, Q factors are assumed to be frequency independent. Simulations showed that changes in the transmission and the reflection are identical for all different topologies from the same polynomial. Hence, the losses are independent from the topologies but dependent on the polynomials.

In the transmission line theory, a real part appears in the phase term to represent the loss. This attenuation constant shifts the complex propagation constant along the real axis. Similarly empirical studies showed that lossless network polynomials' roots are shifted along the real axis of the complex frequency domain when loss is introduced by series resistances. The shift can be calculated as:

$$\alpha_i = -(-1)^k \frac{Im(p_i)}{\sqrt{Q_C Q_L}} , \qquad (5.96)$$

where k is the order of the transmission zero, α_i is a real value, which accounts for the shift in the ith complex root. α_i depends on the roots themselves as well as the Q factors of the inductors and capacitors, which are denoted by p_i, Q_L and Q_C, respectively. k affects the direction of the shift. The roots of the lossy network can be found as:

$$p_{i,lossy} = \alpha_i + p_{i,lossless} . \qquad (5.97)$$

Finally, the polynomials for the lossy networks can be determined when the roots are known.

At a given frequency the shift can be calculated individually for all roots, by which the lossy network can be analyzed. A nth order $h(s)$ polynomial of the lossless network can be written in factors form:

$$h(s)_{lossless} = (s - p_1)(s - p_2) \dots (s - p_n) , \qquad (5.98)$$

where $s = j\omega$ is the natural frequency. When the losses are considered, shifts of the roots and new roots are calculated from Eqn. 5.96 and 5.97, respectively. Then, Eqn. 5.98 can be rewritten for lossy $h(s)$ as following:

$$h(s)_{lossy} = [s - (\alpha_1 + p_1)][s - (\alpha_2 + p_2)] \dots [s - (\alpha_n + p_n)] . \qquad (5.99)$$

For the lossless networks, when a complex frequency $j\omega_0$ is equal to the ith root, the value of the $h(s)$ polynomial is zero. For lossy networks, on the other hand, the value of the polynomial depends on the shift of the ith root since it remains after $j\omega_0$ cancels p_i. Thus, at the specific frequency the shift α_i is enough to estimate the polynomial $h(s)_{lossy}$ while the rest of the shifts can be assumed equal to this α_i. Hence, Eqn. 5.99 can be rewritten for $s = j\omega_0$ as following:

$$h(j\omega_0)_{lossy} = [j\omega_0 - (\alpha_1 + p_1)][j\omega_0 - (\alpha_2 + p_2)]\ldots[j\omega_0 - (\alpha_n + p_n)]$$
$$= h(j\omega_0 - \alpha_i) . \qquad (5.100)$$

The values of the lossy network polynomials can be calculated from the lossless network polynomials by shifting the frequency points. Therefore, this method is called as frequency shifting algorithm (FSA). The closest polynomial root to a given frequency point is determined, and then, the amount of the shift of this root is calculated. The shift is added to the complex frequency in order to analyze the lossy network performance from the lossless network polynomials.

The accuracy of FSA can be verified by an example. In Fig. 5.77, the reflection coefficients of three circuits are compared in a frequency range from DC to 2 GHz. The lossless network is denoted as *Lossless Net*. The second model is called *Lossy Net*, which is obtained when the series resistances are added to the lossless components. Finally another reflection coefficient of the lossy network are determined from modifying lossless polynomials using the FSA, which is denoted as *FSA*.

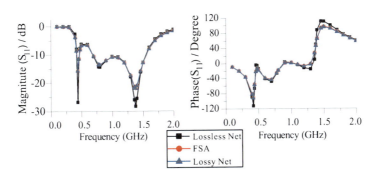

Fig. 5.77 Magnitudes and phases of reflection coefficients of the three circuits. For the lossy networks, Q_C and Q_L are assumed to be 50.

Results for the first and second models are calculated directly on the circuit level. For FSA, the complex roots of the $h(s)$ polynomial of the lossless net are found as $0.003 \pm 0.440j$, $-0.087 \pm 0.781j$ and $-0.023 \pm 1.379j$. Based on the imaginary parts of the roots, the frequency range is divided into three regions, where the imaginary parts are at the middle of the respective regions. By using Eqn. 5.96, the shifts in the regions are calculated as 0.00881, 0.01562 and 0.02759, respectively. In Fig. 5.77, the magnitude and phase of the reflection coefficients obtained from the lossy network and those calculated by FSA matched well.

Since FSA enables to analyze lossy networks on their polynomial level, their performance e.g. insertion loss can be improved as compared to that of the networks designed by conventional way. The conventional way assumes lossless components in the polynomial. Then on the circuit level these components are substituted for available lossy components. Now with FSA the network polynomial is first optimized by taking the components' loss into account. Then despite of the final circuit topology, its optimality is guaranteed. The improvements of both transmission and reflection are depicted in Fig. 5.78. These improvements are not simultaneously achievable. When one of the coefficients is improved the other remains almost the same. It shows that the improvements tend to be increased as the network component number increases or the component Q factor decreases. For practical designs with more than 4 reactive components, the maximal improvement of the reflection is about 15 dB while for the transmission it is about 0.5 dB.

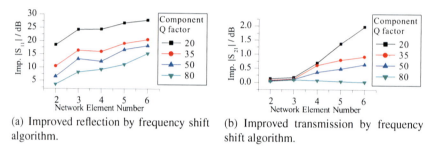

(a) Improved reflection by frequency shift algorithm.

(b) Improved transmission by frequency shift algorithm.

Fig. 5.78 Simulated improvement of reflection and transmission by employing frequency shift algorithm. Here $Q_C = Q_L = Q$.

5.3.3 Impedance Matching Ranges

The above design method consider only fixed frequencies. It is to be shown in the following, that the technique can be extended to design TMN. The target is to enlarge the TMN's conjugate matching range. As in section 5.2.1, at the circuit level, the range can be calculated with the known topology. However, during the design process of MRFT method, the topology is unknown. Therefore, the boundaries of the matching range is to be determined by the polynomials. Afterwards the matching range can be enlarged without having to know the actually implementation topology.

When tuning the capacitors, the conjugate matching range of the TMN has several vertices depending on the tunability and capacitance of the varactors. The exemplary matching range of the topologies in Fig. 5.75 are plotted in the Fig. 5.79 at 0.9 GHz. The topologies B and C have identical matching ranges, which are different from that of topology A.

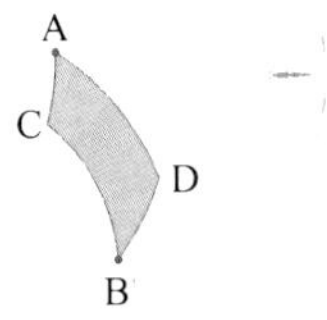

5.3 Tunable Multiband Matching Network

(a) The conjugate matching ranges of topology A in Fig. 5.75(a).

(b) The conjugate matching ranges of the topologies B and C in Fig. 5.75(b) and 5.75(c), respectively.

Fig. 5.79 Simulated matching ranges for various topologies with the same polynomial and the same 30% varactor tunability

These networks have the same Thévenins' impedance polynomial $\bar{Z}_q(s)$, which is given as an example:

$$\bar{Z}_q(s) = \frac{2.000s^3 + 1.6048s^2 + 1.549s + 0.000}{2.257s^4 + 1.811s^3 + 6.144s^2 + 0.990s + 0.9557} . \quad (5.101)$$

Each network has two varactors, which are assumed to have 30% tunability. These varactors have four extreme tuning states. As a result, there are four vertices of the matching range. As shown above, only two of the vertices are common between the three different topologies, which are the vertex A and B, while the other two are different between topologies. Therefore, these points are related with the polynomial set which satisfies the Feldtkeller relation. The tuning states of the capacitors at the vertices are categorized in Table. 5.11.

Table 5.11 Tuning states of matching range vertices

Vertex	Tuning state	Abbreviation	Topology dependent
A	No capacitor is tuned	NTS	Independent
B	All capacitors are tuned	ATS	Independent
C	Shunted capacitors are tuned	ShTS	Dependent
D	Serial capacitors are tuned	SeTS	Dependent

The reason why the NTS and ATS points are common between the topologies can be explained by their Thévenins' impedance polynomials as,

$$Z_A(s) = \frac{s^3(C_1+C_2)L_1L_2 + s^2(C_1+C_2)L_1Z + sL_1}{s^4M + s^3C_1C_2L_1Z + s^2(C_1L_1+C_1L_2+C_2L_2) + sZ(C_1+C_2) + 1}$$

$$Z_B(s) = \frac{s^3C_2L_1L_2 + s^2(C_2L_1+C_2L_2) + s(L_1+L_2)}{s^4M + s^3C_1C_2L_1Z + s^2(C_1L_1+C_1L_2+C_2L_2) + sZC_2 + 1}$$

$$Z_C(s) = \frac{s^3(C_2L_1L_2) + s^2C_2L_1Z + sL_2}{s^4M + s^3C_1C_2L_1Z + s^2(C_1L_1+C_1L_2+C_2L_1) + sZC_2 + 1}, \quad (5.102)$$

where $M = C_1C_2L_1L_2$ and Z is the resistance seen behind the input transformer from the network side.

With the impedances in Eqn. 5.102, the polynomial for each topology is calculated with the corresponding component values. These polynomials are equal to the NTS polynomials of the networks since none of the capacitors are tuned. Thus the networks have the same NTS point for the given complex frequency, namely s. In the ATS state, the capacitors are multiplied by a factor of $(1-\tau)$ where τ is the varactor tunability. For example, for $Z_A(s)$, the multiplication leads to a numerator for the ATS, namely $N_{ATS}(s)$ as:

$$N_{ATS}(s) = s^2(C_1+C_2)L_1L_2(1-\tau) + s^2(C_1C_2)L_1Z(1-\tau) + sL_1. \quad (5.103)$$

If each individual numerator coefficients of $Z_A(s)$ are normalized with the respective NTS numerator coefficients in Eqn. 5.103, the coefficients vector is found:

$$\begin{bmatrix} N_{NTS} \\ N_{ATS} \end{bmatrix} = \begin{bmatrix} \frac{(C_1+C_2)L_1L_2}{(C_1+C_2)L_1L_2(1-\tau)} \\ \frac{(C_1+C_2)L_1Z}{((C_1+C_2)L_1Z(1-\tau)} \\ \frac{L_1}{L_1} \end{bmatrix} = \begin{bmatrix} \frac{1}{1-\tau} \\ \frac{1}{1-\tau} \\ 1 \end{bmatrix}. \quad (5.104)$$

The normalization of the denominator leads to the similar vector:

$$\begin{bmatrix} D_{NTS} \\ D_{ATS} \end{bmatrix} = \begin{bmatrix} \frac{1}{(1-\tau)^2} \\ \frac{1}{(1-\tau)^2} \\ \frac{1}{1-\tau} \\ \frac{1}{1-\tau} \end{bmatrix} \quad (5.105)$$

The right hand sides of the two equations can function as look-up tables, which are determined by the network component number and the number of transmission zeros at the origin. Furthermore, at the same time, the TPG at constant frequencies

5.3 Tunable Multiband Matching Network

can be optimized for both NTS and ATS status at the polynomial level. Therefore, the polynomial level relations of the points enable to increase the distance between each other, resulting in an improvement in the matching range for given number of varactors and their tunability.

5.3.4 Proof of Concept

To prove the above mentioned MRFT method and FSA algorithm, several prototypes are implemented. First, two dual-band static prototypes of same network topology are realized to evaluate the transmission improvement by FSA. The prototype D_1 is designed directly by MRFT method without compensation of component losses. In the other prototype D_2, the components are considered as lossless in MRFT, and then compensated also on the polynomial level by using FSA. Their topology is shown in Fig. 5.80.

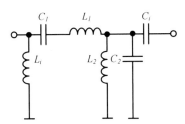

Fig. 5.80 Topology of realized prototypes D_1 and D_2

The component values are given in Table. 5.12. For realization, Coilcraft® 0402CS inductor series and AVX® ACCU-P0402 capacitor series are utilized. Closest components to the theoretical design are chosen. During the D_2 polynomial design, the component Q-factors are taken from the component data sheets and assumed to be constant over the target frequency range, which are 70 and 100 for inductors and capacitors, respectively.

Table 5.12 Components of the prototypes D_1 and D_2. Both designed and realized parameters are listed.

Network	D_1		D_2	
Component	Design	Realization	Design	Realization
L_i(nH)	12.399	12.0	40.744	43.0
L_1(nH)	6.28	6.2	4.081	3.9
L_2(nH)	6.081	6.2	8.17	8.2
C_i(pF)	2.8	2.2	3.38	3.3
C_1(pF)	1.55	1.8	3.73	3.3
C_2(pF)	1.66	1.8	1.59	1.2

Simulated and measured transmission coefficients of D_1 and D_2 are compared in Fig. 5.81. Here the component Q factors are assumed to be frequency independent. In detail the transmission of the D_2 is up to 0.3 dB higher than that of D_1 over the frequency range. It is to conclude that, the component losses are effectively compensated by FSA method.

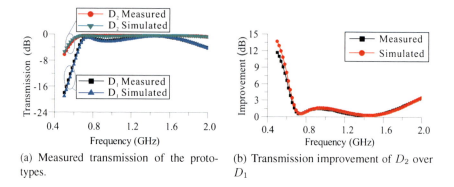

(a) Measured transmission of the prototypes.

(b) Transmission improvement of D_2 over D_1

Fig. 5.81 Measured transmission improvement by employing frequency shift algorithm

Afterwards, a dual-band TMN operating simultaneously at 1 GHz and 1.8 GHz is realized. The prototype is shown in Fig. 5.82. The varactors have a 50 % tunability and a Q factor better than 60. The component values are listed in Table. 5.13. Two 50 Ω ports Anritsu VNA is used for measurements. The bias voltage is generated by an external dual channel voltage source. Fig. 5.83 shows the untuned state. The impedance is well matched across the frequency from 0.8 GHz to 2.6 GHz especially at 1 GHz and 1.8 GHz, while the insertion losses is approximately 0.2 dB.

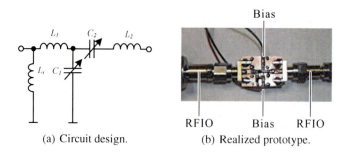

(a) Circuit design.

(b) Realized prototype.

Fig. 5.82 Dual-band tunable matching network prototype designed for simultaneous operation at 1 GHz and 1.8 GHz

5.3 Tunable Multiband Matching Network

Table 5.13 Component valus of the dual-band tunable matching network prototype in Fig. 5.82

L_i	L_1	L_2	C_1	C_2
7.5 nH	1.8 nH	5.6 nH	1.9 pF	2.55 pF

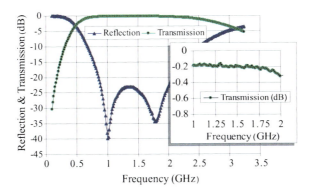

Fig. 5.83 Scattering parameter measurements of the prototype

While tuning the varactors, the corresponding TPG and input reflection are calculated based on the measured scattering parameters. As shown in Fig. 5.84(a) and 5.84(b), the additional loss introduced by the TMN is about 0.1 dB at 1 GHz and 0.35 dB at 1.8 GHz. Meanwhile the reflection coefficients are reduced significantly. For example, at 1 GHz the -4.4 dB reflection domain without the TMN is improved to below -8 dB. In addition the -10 dB reflection domain without the network is enlarged. Especially, the reflection coefficients are reduced for capacitive loads more than those for inductive loads since the TPG is maximized for capacitive impedances at the ATS. These results show that the low insertion losses are obtained with enlarged reflection coefficient range towards the desired type of the load impedances.

It can be concluded that the modified polynomial synthesis method MRFT is efficient in designing multiband TMN. It allows optimization of the networks without any constrains on the exact topology. When incorporating with FSA method, the component losses are minimized. At last, the synthesized networks match the complex load impedance at multiple bands simultaneously, which answers the challenges mentioned at the beginning of this section.

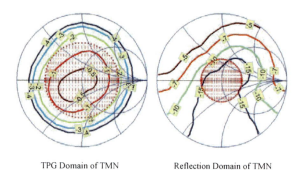

TPG Domain of TMN   Reflection Domain of TMN

(a) Measured performance at 1 GHz. Left, the shaded circle denotes the $-1.93\,dB$ transducer power gain (TPG) domain, which corresponds to $-4.4\,dB$ reflection domain without tunable matching network (TMN). Right, the shaded circle shows the $-10\,dB$ reflection domain without TMN.

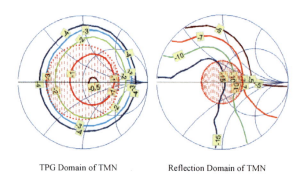

TPG Domain of TMN   Reflection Domain of TMN

(b) Measured performance at 1.8 GHz. Left, the shaded circle denotes the -1.65 dB TPG domain, which corresponds to -5 dB reflection range without TMN. Right, the shaded circle shows the -10 dB reflection range without TMN.

Fig. 5.84 Measured dual band matching performance at 1 GHz and 1.8 GHz, respectively

5.4 Tunable Substrate Integrated Waveguide Bandpass Filter

Electrically tunable filters facilitate the architecture simplification of multiband and wide band wireless systems. Through a dynamical reconfiguration of the operation frequency and bandwidth, they efficiently cope with the time and regional variations of traffic demands. Research has been conducted on various circuit topologies, and the underlying tuning components. Amongst the candidate technologies, the varactors built on ferroelectric materials such as BST exhibit appealing properties at microwave frequencies. They possess applicable tunability, adequate dielectric loss, compactness, low operation energy consumption, high microwave power handling capability, and high tuning speed [83]. Especially, the thick-film ceramics can be fabricated in a low-cost screen printing process. Therefore, they have recently become attractive for the development of tunable resonators, filters, antennas and phase shifters for frequency agile applications.

In the reported tunable bandpass filters with ferroelectric varactors, mainly microstrip or dielectric resonators have been utilized [1, 25, 77]. In the former case, the strong resonance in the comb line and ring structures, as well as the heavy loading of the varactors are designed to guarantee the tuning range, but compromise the passband insertion loss and selectivity. In the latter case, the high Q-factor dielectric resonators and the adequate coupling to shunted varactors allow low insertion loss. However, the dimension of the resonator and the indispensable bulky shielding encumber a compact and integrated packaging. A promising solution towards compact and high Q-factor resonator is the evanescent-mode cavity. A forward waveguide propagation below the cutoff frequency is enabled when the cavity is loaded by reactive scatterers. It is further embodied in substrate integrated waveguide (SIW), which utilizes complementary split-ring resonator (CSRR) as compact scatterer at fixed frequencies [33]. In [39], planar split-ring resonator (SRR) is extended to be tunable by embedding ferroelectric varactors for bandstop applications.

In the following, the realization of compact tunable bandpass filter by introducing tunable CSRR scatterers into a ceramic SIW cavity is investigated. The input impedance matching has been additionally taken into account. A prototype has been realized on top of a BST thick-film ceramic substrate with embedded varactors and integrated bias circuit.

5.4.1 Bandpass Filter Design

The filter's configuration is depicted in Fig. 5.85. It consists of a tunable evanescent-mode SIW cavity, and tunable impedance matching networks. The evanescent-mode SIW cavity is loaded with a pair of tunable mushroom-type CSRRs.

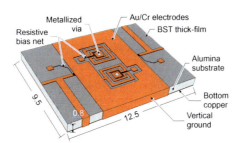

Fig. 5.85 Overview of the proposed tunable filter. Circuitries are implemented on the top of a ceramic substrate. The substrate integrated waveguide cavity and microstrip inductors are grounded at board edges. All dimensions are in millimeter unit.

Each matching network is formed by a serial varactor and a shunted microstrip inductor. The whole module is based on an aluminum oxide ceramic substrate with a Cu-doped BST thick-film screen printed on top.

5.4.2 Tunable Evanescent-Mode Substrate Integrated Waveguide Cavity

The rectangular evanescent-mode SIW cavity is embedded in the aluminum oxide substrate with a relative permittivity of 9.8. Without the scatterers, the cutoff frequency of fundamental TE_{10}-mode is 5.05 GHz. When operating below the cutoff frequency, a forward propagation can be sustained in two means. First, the passband of the quasi-TE_{10}-mode can be shifted lower by adding capacitive ridges in the E-plane. Second, by introducing resonant scatterers, a narrow passband can be obtained, while the TE_{10}-mode is suppressed [33]. As in Fig. 5.86, following the second concept with a pair of oppositely oriented CSRR scatterers, the cavity has two transmission poles, which are resembled respectively by the differential and the common resonate modes of the CSRR pair. For the purpose of minimization, the planar CSRR is extended to mushroom-type as in [73], where the central patch is grounded through a via. Varactors are further introduced between CSRR's inner ring and the central patch.

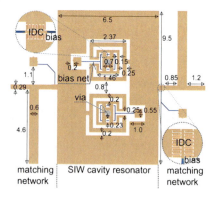

Fig. 5.86 Optimized dimensions of the proposed tunable filter. All are in mm unit.

In each CSRR there are four IDC pairs at vertexes of the patch. The IDCs are built on the BST thick-film, by applying external electrostatic field across the gaps between digits, the capacitance varies, which tunes the CSRRs. The IDC pair is biased at the middle through a highly resistive line. The integrated topology decouples RF signal and external electrostatic bias, without compromising the overall tunability and Q factor as measured and depicted in Fig. 5.87. In order to deliver DC bias from contact pads to IDCs, the CSRR is split by a 0.25 mm-wide gap at the outer ring to accommodate the resistive line, which has negligible disturbance to the symmetric current distribution along the gap.

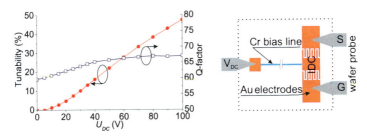

Fig. 5.87 Measured tunability and quality factor of the varactor pair at 3 GHz. The varactor pair is biased at the middle through a resistive line and measured by a signal-ground probe at the gold electrodes.

A fullwave simulation has been performed in CST Microwave Studio®. As shown in Fig. 5.88, if varying the IDCs between 0.3 pF and 1.2 pF, the frequencies of the resonate modes can be shifted from 3.59 GHz to 2.54 GHz and from 3.75 GHz to 2.65 GHz for common and differential mode, respectively. When compared to the mushroom-type CSRR without lumped capacitor loading, the footprint of the tunable CSRR is further minimized by 17 % to 41 %.

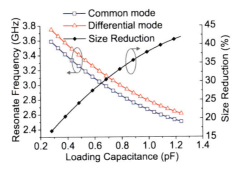

Fig. 5.88 Simulated tuning of the resonance frequencies of the common and differential modes. The complementary split-ring resonator is equivalently minimized in the mean time.

5.4.3 Tunable Impedance Matching Network

When the SIW cavity is tuned across the above mentioned frequency range, its input impedance alters from $26.5 + 44.1i\,\Omega$ at low end to $64.7 + 35.6i\,\Omega$ at high end. The mismatch to $50\,\Omega$ external ports compromises the transmission. A pair of tunable impedance matching networks are implemented right at the ports of SIW cavity, as depicted in Fig. 5.86. Each matching network consists of an IDC varactor at the input port of SIW cavity and a shunted microstrip inductor. As illustrated in Fig. 5.89, the matching network introduces the control over both insertion loss (IL) and fractional bandwidth (FBW). For a given frequency tuned by the CSRR's varactors, the capacitance for the matching network can be then optimized.

An equivalent circuit of the whole module at 2.95 GHz is depicted in Fig. 5.90. The varactors namely R_{vl} and R_{vr} load equivalently in parallel to the CSRR's intrinsic reactance. Both input reflection and transmission coefficients are used in a curve fitting at the frequency of interest. The fitted component values are listed in Table. 5.14.

Table 5.14 Component values of equivalent circuit in Fig. 5.90. All the inductors are in nH unit, capacitors in pF, resistors in Ω, and lengths in mm.

L_{sp}	L_c	L_r	L_v	L_s	L_g	l_p	l_m	l_k
1.25	1.3	1.38	0.35	1.95	0.90	0.35	3.6	0.8

C_c	C_r	C_s	Z_p	Z_k	R_r	R_v	R_c
1.7	0.29	1.65	68	7.15	0.21	0.02	0.45

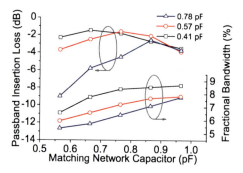

Fig. 5.89 Simulated reduction of insertion loss under condition of different IDC capacitances in CSRR, when using tunable matching networks. The fractional bandwidth alters when the coupling to the cavity changes.

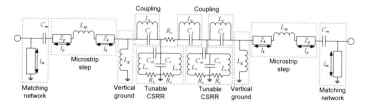

Fig. 5.90 Equivalent circuit of the filter. The varactors are denoted as C_{vl}, C_{vr} and C_m, respectively.

5.4.4 Realization and Measurement

With the above mentioned parameters, a prototype is realized on a 650 μm-thick Al_2O_3 substrate with 2.8 μm BST thick-film screen printed on top. The BST-layer exhibits a relative permittivity of 416 at 3 GHz and room-temperature, with a loss tangent of 0.014. 0.3 mm-diameter vias are drilled through substrate using laser. They are afterwards metalized using conductive polymer ProConduct® from LPKF. The vias introduce about 0.02 Ω resistance denoted as R_v in Fig. 5.90. A thin chromium and gold seed layer is evaporated above. Then the IDCs are realized with 3.1 μm-thick plated gold electrodes. Additional 4.4 μm-thick gold is plated on patches and rings, which in total is more than 5 times the skin depth at 3 GHz to reduce metallic loss. 20 μm-wide and approximately 30 nm-thick bias lines are etched on the chromium seed layer, which show 6 kΩ/mm resistivity per line length. The realized CSRR is depicted in Fig. 5.91. A piece of copper sheet is attached to the substrate's bottom. The vertical groundings are finally realized using the conductive polymer.

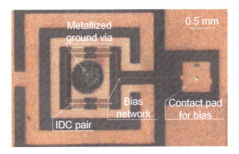

Fig. 5.91 The realized planar tunable complementary split-ring resonator with a ground via at center and four varactor pairs at vertexes. Bias network is resembled by resistive chromium strips in dark color.

The untuned capacitance of each varactor pair in CSRR is 0.39 pF. With 100 V bias voltage across the 6 μm gap, 47 % tunability and Q-factor above 60 are achieved in the frequency range. The varactors in matching networks are 0.86 pF for operation at 2.95 GHz, and then, tuned to 0.5 pF for 3.57 GHz. The test fixture is shown in Fig. 5.92.

The transmission and reflection coefficients are measured during tuning as shown in Fig. 5.93. The center frequency of passband is tunable from 2.95 GHz to 3.57 GHz, i.e. 21 % tunability. The 3 dB fractional bandwidth is below 5.4 %. The reflection is lower than -10.9 dB at the passband center, while the insertion loss is between 3.3 dB and 2.6 dB. Besides the metallic loss of the microstrip structures, the insertion loss is raised considerably by the loss in BST varactors and the inductive rings. The varactors' equivalent series resistance is up to 1.1 Ω, and it is 0.21 Ω of the rings. According to post-simulations, they are expected to introduce 1.4 dB and 0.5 dB loss, respectively, at the low frequency end.

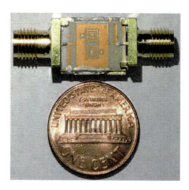

Fig. 5.92 The prototype built on ceramic substrate. The whole module is in a test fixture with SMA connectors.

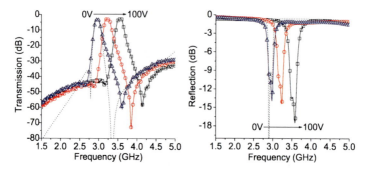

Fig. 5.93 S-parameters of the prototype. Measured: solid lines, simulated at 0 V bias: dot lines.

A comparison with related designs is summarized in Table. 5.15. As shown, the proposed filter exhibits a promising balance between compactness and performance, including frequency coverage and insertion loss. The filter supports the continuous tuning across neighboring multiple bands. It can be later extended in the form of a tunable duplexer by connecting two of them in parallel with a buffer band in between.

Table 5.15 Comparison between the proposed filter and references

Ref.	Size (λ_0*)	Range (GHz)	FBW (%)	IL (dB)
[77]	$0.35 \times 0.35 \times 0.008$	$5.1 - 5.5$	$\approx 6.0^+$	$6.1 - 4.5$
[25]	$0.18 \times 0.10 \times 0.001$	$7.8 - 8.65$	$10^!$	$4.4 - 2.3$
[1]	$0.36 \times 0.36 \times 0.25$	$10.5 - 10.9$	$1.2^!$	$3.5 - 2.0$
this work	$0.10 \times 0.14 \times 0.008$	$2.95 - 3.57$	5.4^+	$3.3 - 2.6$

* normalized to the free-space wavelength at center of the frequency range
+ 3 dB fractional bandwidth
! 1 dB fractional bandwidth

Chapter 6
Summary and Outlook

In this work, the investigation across micro processing technologies and efficient design methodologies has led to tunable multiband ferroelectric devices with improved performance and novel functionalities. It reveals the potential of implementing such devices in reconfigurable RF frontends. The ferroelectric thin-film and thick-film provide the devices with sufficient tunability to cope with services' variation and environmental impact. Meanwhile, the affordable dielectric loss guarantees the devices' efficiency which can be further compensated with optimized designs.

By utilizing the ferroelectric components such as BST varactors, the wireless link including multiband tunable antenna, tunable matching network, and tunable substrate integrated waveguide filter is addressed. The challenging trend towards compactness with wider spectrum coverage is coped with several tunable resonant modes of the antennas with integrated BST varactors. On one hand, such antennas provide narrower stationary bandwidth than the static counterparts. On the other hand, they can be tuned to the desired frequency to equivalently increase the spectrum coverage. A detailed theoretical analysis quantifies that the dynamic fractional bandwidth of the tunable antenna always increases, when the varactor tunability is larger than the stationary fractional bandwidth of the unloaded static antenna. It provides a simple but referential criteria for both the antenna design and the screening of tunable device technologies. A capacitively loaded antenna is modeled for its nonlinearity. As expected, by using ferroelectric varactors instead of diodes, the harmonic radiation at high operation power is suppressed by 35.7 dB at -10 dBm input. A dual-band tunable antenna is presented, which simultaneously covers two bands from 705 MHz to 957 MHz and from 1.4 GHz to 2.25 GHz, respectively. The operation frequency of each band is independently controllable. The concepts are finally extended to a fully integrated ceramic antenna, specifically optimized for frequency division duplex services. A prototype covers the frequency range from 1.47 GHz to 1.76 GHz with a variable distance between the up- and down-link channels from 38 MHz to 181 MHz. The antenna consists of two adjacent slots, both of which are loaded with ferroelectric varactors. They are biased through integrated resistive strip lines. The antenna has a small footprint. When looking forward, by combining such reconfigurability with compact antennas, e.g. bulk ceramic loaded antennas, the pursuit towards antenna compactness as well as wider frequency coverage may be efficiently sustained.

The environmental impact and frequency dependence of antennas can be compensated by the tunable matching network based on BST varactors. The theoretical constraints on the performance of tunable matching networks with lossy varactors and static inductors are identified at single and multiple frequencies. A complete design method for single band matching networks is proposed, and afterwards extended to multiband designs. In these procedures, varactor loss is considered in the circuit and transmission polynomial level, which help to efficiently reduce the insertion loss. Demonstrators have been developed within the BMBF funded MARIO-Project (Multi Access System in Package Radio). With an exemplary antenna exhibiting a reflection higher than -7.5 dB, such a demonstrator considerably mitigates the reflection to below -15 dB across the frequency range from 1.5 GHz to 1.8 GHz. As a contribution to the research project, the study of the design, realization and integration technologies has led to the first integrated ferroelectric thick-film tunable matching network at 2 GHz. It is a multilayer module within a 3 mm × 3 mm package, exhibiting an insertion loss between 0.86 dB and 0.98 dB. The module can efficiently stabilize the antenna impedance across the frequency range from 1.8 GHz to 2.1 GHz. The study of efficient adaptive control concludes that a 6 bit DAC is sufficient for a stable improvement of transducer gain. Future steps towards multiband ferroelectric tunable matching network would involve the development of dense arrays of BST varactors with low RF mutual coupling, independent DC biasing, reliable connection, etc. Furthermore, the merge of tunable antennas and tunable matching networks, both based on ferroelectric components, shall lead to novel low-cost antenna modules, which may be easier to deploy and more persistent to environmental impact.

To enable a bandpass filter, an evanescent mode substrate integrated waveguide cavity in ferroelectric ceramic substrate is loaded with a pair of complementary split ring resonators. The ring resonators are tuned through embedded varactors. The input mismatch due to impedance drift during tuning is compensated by integrated tunable matching networks with varactors. A prototype has been realized in a 12.5 mm × 9.5 mm × 0.8 mm planar module. It shows a frequency coverage from 2.95 GHz to 3.57 GHz, a 3 dB fractional bandwidth of up to 5.4 %, with a comparatively low insertion loss between 3.3 dB and 2.6 dB.

Novel basic devices with BST thin-film are proposed. The multilayer thin-film technology is extended to accommodate additional layers. By introducing a 20 nm thin conductive layer in the middle of two stacked BST films, then applying antipolarized electric fields to the two layers of BST film, the unavoidable generation of acoustic resonance due to the electrostriction in the two layers is considerably suppressed. The method shows the potential to widen the applicable frequency range of BST thin-film varactors. However, there is a reduction of Q factor due to the insertion of lossy middle electrode. It might be improved by combining metallic electrode and conductive oxides bias lines. Furthermore, a controllable electron injection is introduced by the insertion of a 2.5 nm to 10 nm thin highly resistive aluminum oxide layer at the interface between metallic electrode and BST thin-film.

It allows a controllable charge storage at the interface between the alumina layer and the BST film. Once injected, the charge biases the BST thin-film without external electrostatic field until it is reversely discharged. Hence the capacitance can switch between two states which forms a memory window. In an extrapolation, the capacitor holds the memory window up to 15 % in 1 year, and reduces to 6 % in 10 years. It turns to be a functional programmable bi-stable capacitor. More importantly, a low loss capacitor is achievable in contrast to the capacitors with BST in ferroelectric phase, since the thin-film is in low loss paraelectric phase here. The novel bi-stable high frequency capacitor is proposed for the first time. It has inspired further interdisciplinary research to understand deeper the mechanism and utilize more efficiently the induced hysteresis in wireless applications like RFID backscatters.

Appendix A
Technology Parameters

Table A.1 Metallization Deposition Parameters

Compounds	Deposition Method	Rate
Au	thermal evaporation	$11\,nm\,\text{min}^{-1}$
	galvanization	$\leq 100\,nm\,\text{min}^{-1}$ ($\leq 20\,\mu A\,\text{mm}^{-2}$)
Cr	thermal evaporation	$10\,nm\,\text{min}^{-1}$
Ti	electron beam	$1\,nm\,s^{-1}$ to $3\,nm\,s^{-1}$
ITO	electron beam	$3\,nm\,s^{-1}$ to $5\,nm\,s^{-1}$
Pt	RF-sputtering	$2\,nm\,s^{-1}$ to $20\,nm\,s^{-1}$

Table A.2 Etchant Solutions

Compounds	Formula	Quantity
Au etchant		
Iodine	I_2	5 g
Potassium iodide	KI	20 g
Deionized water	H_2O	200 ml
Cr etchant		
Cer-ammonium-nitrate	$(NH_4)_2[Ce(NO_3)_6]$	41.15 g
Nitric acid	HNO_3	22.5 ml
Deionized water	H_2O	250 ml
Ti etchant		
Ammonium fluoride	NH_4F	50 g
Hydrofluoric acid	HF	8 ml
Deionized water	H_2O	200 ml
BST etchant		
Ammonium fluoride	NH_4F	15 g
Hydrofluoric acid	HF	3 ml
Deionized water	H_2O	80 ml
Nitric acid	HNO_3	4 ml

Table A.3 Parameters of RIE

Gas	Pressure	Flow	Power
O_2	150 mTorr	100 cm^3 min^{-1}	300 W
N_2	150 mTorr	20 cm^3 min^{-1}	250 W

Table A.4 Etching Parameters

Compounds	Etching Method	Rate
Au	Wet	3 nm s^{-1}
Cr	Wet	0.5 nm s^{-1}
Ti	Wet	2 nm s^{-1}
BST	Wet	11 nm s^{-1} to 15 nm s^{-1}
	Ar RIE	2 nm min^{-1} to 3 nm min^{-1}
SiO_2	Wet	0.8 nm s^{-1} to 1.1 nm s^{-1}
Si	Ar RIE	6 nm min^{-1} to 10 nm min^{-1}

Table A.5 Parameters of Laser Drilling

Parameters	Value
Voltage	300 V
Duration	0.4 ms
Frequency	100 Hz
Feed speed	50 mm min^{-1}
Nitrogen pressure	10 bar

References

[1] Alford, N., Buslov, O., Keis, V., Kozyrev, A., Petrov, P., Shimko, A.: Band-pass tunable ferroelectric filter based on uniplanar dielectric resonators. In: 38th European Microwave Conference, EuMC 2008, pp. 1703–1706 (2008)

[2] Baniecki, J.D., Shioga, T., Kurihara, K.: Microstructural and electrical properties of $(Ba_xSr_{1-x})Ti_{1+y}O_{3+z}$ thin films prepared by RF magnetron sputtering. Integrated Ferroelectrics 46, 221–232 (2002)

[3] Bartia, P., Bahl, I.: Frequency agile microstrip antennas. Microwave Journal 37, 1136–1139 (1982)

[4] Basceri, C., Streiffer, S.K., Kingon, A.I., Waser, R.: The dielectric response as a function of temperature and film thickness of fiber-textured $(Ba,Sr)TiO_3$ thin films grown by chemical vapor deposition. Journal of Applied Physics 82, 2497–2504 (1997)

[5] Baumert, B.A., Chang, L.H., Matsuda, A.T., Tsai, T.L., Tracy, C.J., Gregory, R.B., Fejes, P.L., Cave, N.G., Taylor, D.J., Otsuki, T., Fujii, E., Hayashi, S., Suu, K.: Characterization of sputtered barium strontium titanate and strontium titanate thin films. Journal of Applied Physics 82, 2558–2566 (1997)

[6] Behdad, N., Sarabandi, K.: Dual-band reconfigurable antenna with a very wide tunability range. IEEE Transactions on Antennas and Propagation 54(2), pt. 1, 409–416 (2006)

[7] Behdad, N., Sarabandi, K.: A varactor-tuned dual-band slot antenna. IEEE Transactions on Antennas and Propagation 54(2), pt. 1, 401–408 (2006)

[8] Behrisch, R.: Sputtering by Particle bombardment. Springer, Berlin (1981)

[9] Belevitch, V.: Classical network theory. Holden Day, San Francisco (1968)

[10] Black, C.T., Welser, J.J.: Electric-field penetration into metals: Consequences for high-dielectric-constant capacitors. IEEE Transactions on Electron Devices 46, 776–780 (1999)

[11] Brendan, K.P.S.: Principles of Dielectrics, 2nd edn. Oxford University Press (September 3, 1998)

[12] Carlin, H.: A new approach to gain-bandwidth problems. IEEE Transacations on Circuits System CAS-24(4) (1977)

[13] Carlin, H.: A new approach to gain bandwidth problem. IEEE Transacations on Circuits and System 23, 170–175 (1957)

[14] Carlin, H.: On optimum broad-band matching. IEEE Transactions on Circuits and System 28, 401–405 (1981)

[15] Carlin, H., Komiak, J.: A new method of broad-band equalization applied to microwave amplifiers. IEEE Transactions on Microwave Theory and Techniques MTT-27(2) (February 1979)

[16] Carlin, H.J.: Darlington synthesis revisited. IEEE Transacations on Circuits System: Fundamental Theory and Applications 46(1), 14–21 (1999)

[17] Carlin, H.J., Yarman, B.S.: The double matching problem: Analytic and real frequency solutions. IEEE Transacations on Circuits System 30(1) (January 1983)

[18] Chang, W.T., Horwitz, J.S., Carter, A.C., Pond, J.M., Kirchoefer, S.W., Gilmore, C.M., Chrisey, D.B.: The effect of annealing on the microwave properties of $Ba_{0.5}Sr_{0.5}TiO_3$ thin films. Applied Physics Letters 74, 1033–1035 (1999)

[19] Chen, L.-Y.V., Forse, R., Chase, D., York, R.: Analog tunable matching network using integrated thin-film BST capacitors. IEEE Microwave Symposium Digest 1, 261–264 (2004)

[20] Chrisey, D.B., Hubler, G.K.: Pulsed Laser Deposition of Thin Films. John Wiley & Sons (1994)

[21] Chu, L.J.: Physical limitations of omni-directional antennas. Applied Physics 19, 1163–1175 (1948)

[22] Cohn, M., Eikenberg, A.: Ferroelectric phase shifters for VHF and UHF. IRE Transactions on Microwave Theory and Techniques 10(6), 536–548 (1962)

[23] Coleman, T., Li, Y.: An interior, trust region approach for nonlinear minimization subject to bounds. SIAM Journal on Optimization 6, 418–445 (1996)

[24] Collin, R.E., Rothschild, S.: Evaluation of antenna q. IEEE Transactions on Antennas and Propagation 12, 23–27 (1964)

[25] Courreges, S., Lacroix, B., Amadjikpe, A., Phillips, S., Zhao, Z., Choi, K., Hunt, A., Papapolymerou, J.: Back-to-back tunable ferroelectric resonator filters on flexible organic substrates. IEEE Transactions on Ultrasonics, Ferroelectrics and Frequency Control 57(6), 1267–1275 (2010)

[26] Cross, L., Newnham, R.E.: History of Ferroelectrics. In: Ceramics and Civilization III: High-Technology Ceramics–Past, Present and Future, pp. 289–305. The American Ceramic Society, Westerville (1988)

[27] Das, S.N.: Quality of a ferroelectric material. IEEE Transactions on Microwave Theory and Techniques 12(7), 440 (1964)

[28] Dauguet, S., Gillard, R., Citerne, J., Piton, G.: Global electromagnetic analysis of microstrip agile antenna. Electronics Letters 33(13), 1111–1112 (1997)

[29] Boucher, D., Lagier, M., Maerfeld, C.: Computation of the vibrational modes for piezoelectric array transducers using a mixed finite element-perturbation method. IEEE Trans. on Sonic. and Ultrasonics SU-28(5), 318–330 (1981)

[30] Dedieu, H., Dehollain, C., Neirynck, J., Rhodes, G.: A new method for solving broadband matching problems. IEEE Transactions on Fundamental Theory and Applications 41, 561–571 (1994)

[31] Desoli, G., Filippi, E.: An outlook on the evolution of mobile terminals. IEEE Circuits Syst. Mag. 6(2) (2006)

[32] Domenico, M.D., Johnson, D.A., Pantell, R.H.: Ferroelectric harmonic generator and the large-signal microwave characteristics of a ferroelectric ceramic. Journal of Applied Physics 33, 1697 (1962)

[33] Dong, Y.D., Yang, T., Itoh, T.: Substrate integrated waveguide loaded by complementary split-ring resonators and its applications to miniaturized waveguide filters. IEEE Transactions on Microwave Theory and Techniques 57(9), 2211–2223 (2009)

[34] Ellmer, K.: Magnetron sputtering of transparent conductive zinc oxide: relation between the sputtering parameters and the electronic properties. Journal of Physics D - Applied Physics 33, R17–R32 (2000)
[35] Fano, R.M.: Theoretical limitations on the broad-band matching of arbitrary impedances. Jouranl of the Franklin Institue 249(1,2), 57–83, 139–154 (1950)
[36] Gautschi, G.: Piezoelectric Sensorics: Force, Strain, Pressure, Acceleration and Acoustic Emission Sensors, Materials and Amplifiers. Springer (2002)
[37] Gevorgian, S.: Ferroelectrics in Microwave Devices. Springer (2009)
[38] Gevorgian, S., Vorobiev, A., Lewin, T.: DC field and temperature dependent acoustic resonances in parallel-plate capacitors based on SrTiO3 and Ba0.25Sr0.75TiO3 films - experiment and modeling. Journal of Applied Physics 99, 124112-11 (2006)
[39] Gil, M., Damm, C., Giere, A., Sazegar, M., Bonache, J., Jakoby, R., Martin, F.: Electrically tunable split-ring resonators at microwave frequencies based on barium-strontium-titanate thick films. Electronics Letters 45(8), 417–418 (2009)
[40] Gurevich, V.L., Tagantsev, A.K.: Intrinsic dielectric loss in crystals. Advances in Physics 40, 719–767 (1991)
[41] Gustafsson, M., Nordebo, S.: Bandwidth, Q factor, and resonance models of antennas. Progress In Electromagnetics Research 62, 1–20 (2006)
[42] Hall, P., Kapoulas, S., Chauhan, R., Kalialakis, C.: Microstrip antennas with adaptive integrated tuning. In: Proc. Int. Conf. on Antennas and Propagation, ICAP 1997, vol. 1, pp. 501–504 (1997)
[43] Hao, L.Z., Zhu, J., Luo, W.B., Zeng, H.Z., Li, Y.R., Zhang, Y.: Electron trap memory characteristics of LiNbO$_3$ film/AlGaN/GaN heterostructure. Applied Physics Letters 96, 032103 (2010)
[44] Heinen, S.: High dynamic range rf frontends from multiband multistandard to cognitive radio. In: Semiconductor Conference Dresden, SCD (2011)
[45] Horikawa, T., Mikami, N., Makita, T., Tanimura, J., Kataoka, M., Sato, K., Nunoshita, M.: Dielectric-Properties of (Ba, Sr)TiO$_3$ Thin-Films Deposited by Rf-Sputtering. Japanese Journal of Applied Physics Part 1 - Regular Papers Short Notes & Review Papers 32, 4126–4130 (1993)
[46] Im, J., Auciello, O., Baumann, P.K., Streiffer, S.K., Kaufman, D.Y., Krauss, A.R.: Composition-control of magnetron-sputter-deposited $(Ba_xSr_{1-x})Ti_{1+y}O_{3+z}$ thin films for voltage tunable devices. Applied Physics Letters 76, 625–627 (2000)
[47] Janaswamy, R., Schaubert, D.H.: Characteristic impedance of a wide slotline on low permittivity substrates. IEEE Transactions on Microwave Theory and Techniques MTT-34, 900–902 (1986)
[48] Jeon, J.H.: Effect of SrTiO$_3$ concentration and sintering temperature on microstructure and dielectric constant of $Ba_{1-x}Sr_xTiO_3$. Journal of the European Ceramic Society 24, 1045–1048 (2004)
[49] Johnson, D.H.: Origins of the equivalent circuit concept: the voltage-source equivalent. Proceedings of the IEEE 91(4), 636–640 (2003)
[50] Jondral, F.: Software-defined radio - basics and evolution to cognitive radio. EURASIP J. Wireless Comm. Network. 3, 275–283 (2005)
[51] Kalialakis, C., Hall, P.: Analysis and experiment on harmonic radiation and frequency tuning of varactor-loaded microstrip antennas. IET Microwaves, Antennas & Propagation 1(2), 527–535 (2007)
[52] Kittel, C.: Introduction to solid state physics, 4th edn. John Wiley & Sons, Inc., New York (1996)

[53] Ko, S., Murch, R.: A diversity antenna for external mounting on wireless handsets. IEEE Transactions on Antennas and Propagation 49(5), 840–842 (2001)
[54] Lakin, K., Kline, G., McCarron, K.: High-q microwave acoustic resonators and filters. IEEE Transactions on Microwave Theory and Techniques 41(12), 2139–2146 (1993)
[55] Lampert, M.A., Mark, P. (eds.): Current injection in solids. Academic Press, New York (1970)
[56] Lee, C.H., Hur, S.H., Shin, Y.C., Choi, J.H., Park, D.G., Kim, K.: Charge-trapping device structure of SiO_2/SiN/high-k dielectric Al_2O_3 for high-density flash memory. Applied Physics Letters 86 (2005)
[57] Lee, W.J., Kim, H.G., Yoon, S.G.: Microstructure dependence of electrical properties of $(Ba_{0.5}Sr_{0.5})TiO_3$ thin films deposited on $Pt/SiO_2/Si$. Journal of Applied Physics 80, 5891–5894 (1996)
[58] Li, S.Y., Wachau, A., Schafranek, R., Klein, Y.L.Z.A., Jakoby, R.: Energy level alignment and electrical properties of $(Ba, Sr)TiO_3/Al_2O_3$ interfaces for tunable capacitors. Journal of Applied Physics 108, 014113 (2010)
[59] Lide, D.R.: Handbook of Chemistry and Physics, 90th edn. CRC Press (June 3, 2009)
[60] Lo, S.H., Buchanan, D.A., Taur, Y., Wang, W.: Quantum-mechanical modeling of electron tunneling current from the inversion layer of ultra-thin-oxide nmosfet's. IEEE Electron Device Letters 18, 209–211 (1997)
[61] Lyddane, R.H., Sachs, R.G., Teller, E.: On the polar vibrations of alkali halides. Physical Review 59, 673 (1941)
[62] Dillinger, K.M.M., Alonistioti, N.: Software Defined Radio: Architectures, Systems and Functions. Wiley, Chichester (2003)
[63] Maikap, S., Lee, H.Y., Wang, T.Y., Tzeng, P.J., Wang, C.C., Lee, L.S., Liu, K.C., Yang, J.R., Tsai, M.J.: Charge trapping characteristics of atomic-layer-deposited HfO_2 films with Al_2O_3 as a blocking oxide for high-density non-volatile memory device applications. Semiconductor Science and Technology 22, 884–889 (2007)
[64] Mandel, C., Maune, H., Maasch, M., Sazegar, M., Schuessler, M., Jakoby, R.: Passive wireless temperature sensing with BST-based chipless transponder. In: German Microwave Conference, GeMIC (2011)
[65] Nakamura, K., Kobayashi, H., Kanbara, H.: Evaluation of acoustic properties of thin films using piezoelectricovertone thickness-mode resonators. In: IEEE Ultrasonics Symposium, vol. 1, pp. 593–597 (2000)
[66] Nakata, S., Nagai, S., Kumeda, M., Kawae, T., Morimoto, A., Shimizu, T.: Etching rate, optical transmittance, and charge trapping characteristics of Al-rich Al_2O_3 thin film fabricated by rf magnetron cosputtering. Journal of Vacuum Science & Technology B 26, 1373–1378 (2008)
[67] Natori, K., Otani, D., Sano, N.: Thickness dependence of the effective dielectric constant in a thin film capacitor. Applied Physics Letters 73, 632–634 (1998)
[68] Nespurek, S., Zmeskal, O., Sworakowski, J.: Space-charge-limited currents in organic films: Some open problems. Thin Solid Films 516, 8949–8962 (2008)
[69] Oh, U.C., Kang, T.S., Park, K.H., Je, J.H.: Effects of plasma power on the epitaxial growth of $Ba_{0.48}Sr_{0.52}TiO_3$ thin film. Journal of Applied Physics 86, 163–167 (1999)
[70] Ohring, M.: Materials Science of Thin Films: Deposition & Structure, 2nd edn. Academic Press, San Diego (2002)
[71] Panayi, P., Al-Nuaimi, M., Ivrissimtzis, I.: Tuning techniques for planar inverted-f antenna. Electronics Letters 37(16), 1003–1004 (2001)
[72] Patnaik, P. (ed.): Handbook of Inorganic Chemicals. McGraw-Hill (2002)

References

[73] Peng, L., Ruan, C.-L., Li, Z.-Q.: A novel compact and polarization-dependent mushroom-type EBG using CSRR for dual/triple-band applications. IEEE Microwave and Wireless Components Letters 20(9), 489–491 (2010)

[74] Pertsev, N.A., Zembilgotov, A.G., Hoffmann, S., Waser, R., Tagantsev, A.K.: Ferroelectric thin films grown on tensile substrates: Renormalization of the curie-weiss law and apparent absence of ferroelectricity. Journal of Applied Physics 85, 1698–1701 (1999)

[75] Pervez, N.K., Hansen, P.J., York, R.A.: High tunability barium strontium titanate thin films for rf circuit applications. Applied Physics Letters 85, 4451–4453 (2004)

[76] Pijolat, M., Loubriat, S., Mercier, D., Reinhardt, A., Defay, E., Deguet, C., Aid, M., Queste, S., Ballandras, S.: LiNbO$_3$ film bulk acoustic resonator. In: 2010 IEEE International Frequency Control Symposium (FCS), vol. 661-664 (2010)

[77] Pleskachev, V., Vendik, I.: Tunable microwave filters based on ferroelectric capacitors. In: 15th International Conference on Microwaves, Radar and Wireless Communications, MIKON 2004, vol. 3, pp. 1039–1043 (2004)

[78] Pozar, D.M.: Microwave Engineering. John Wiley & Sons, New York (1998)

[79] Reinhardt, A., Ballandras, S., Laude, V.: Simulation of transverse effects in fbar devices. In: 2005 IEEE MTT-S International Microwave Symposium Digest (2005)

[80] Rostbakken, O., Hilton, G., Railton, C.: Adaptive feedback frequency tuning for microstrip patch antennas. In: Proc. Int. Conf. on Antennas and Propagation, pp. 166–170 (1995)

[81] Schade, K., Suchaneck, G., Tiller, H.J.: Plasma Technik: Anwendung in der Elektronik. Verlag Technik, Berlin (1990)

[82] Schafranek, R., Giere, A., Balogh, A.G., Enza, T., Zheng, Y., Scheele, P., Jakoby, R., Klein, A.: Influence of sputter deposition parameters on the properties of tunable barium strontium titanate thin films for microwave applications. Journal of the European Ceramic Society 29, 1433–1442 (2007)

[83] Scheele, P., Giere, A., Zheng, Y., Goelden, F., Jakoby, R.: Modeling and applications of ferroelectric-thick film devices with resistive electrodes for linearity improvement and tuning-voltage reduction. IEEE Transactions on Microwave Theory and Techniques 55(2), pt. 2, 383–390 (2007)

[84] Scheele, P., Goelden, F., Giere, A., Mueller, S., Jakoby, R.: Continuously tunable impedance matching network using ferroelectric varactors. In: IEEE Microwave Symposium Digest (2005)

[85] Schmidt, M., Lourandakis, E., Leidl, A., Seitz, S., Weigel, R.: A comparison of tunable ferroelectric Π- and T-matching network. In: Proceedings of European Microwave Conference, pp. 98–101 (2007)

[86] Shaw, T.M., Suo, Z., Huang, M., Liniger, E., Laibowitz, R.B., Baniecki, J.D.: The effect of stress on the dielectric properties of barium strontium titanate thin films. Applied Physics Letters 75, 2129–2131 (1999)

[87] Shih, W.K., Wang, E.X., Jallepalli, S., Leon, F., Maziar, C.M., Taschjr, A.F.: Modeling gate leakage current in nmos structures due to tunneling through an ultra-thin oxide. Solid-State Electronics 42, 997–1006 (1998)

[88] Simmons, J.G.: Richardson-schottky effect in solids. Physical Review Letters 15, 967 (1965)

[89] Sinnamon, L.J., Bowman, R.M., Gregg, J.M.: Investigation of dead-layer thickness in SrRuO$_3$/Ba$_{0.5}$Sr$_{0.5}$TiO$_3$/Au thin-film capacitors. Applied Physics Letters 78, 1724–1726 (2001)

[90] Sinnamon, L.J., Saad, M.M., Bowman, R.M., Gregg, J.M.: Exploring grain size as a cause for "dead-layer" effects in thin film capacitors. Applied Physics Letters 81, 703–705 (2002)

[91] Specht, M., Reisinger, H., Hofmann, F., Schulz, T., Landgraf, E., Luyken, R.J., Roesner, W., Grieb, M., Risch, L.: Charge trapping memory structures with Al_2O_3 trapping dielectric for high-temperature applications. Solid-State Electronics 49, 716–720 (2005)

[92] Specht, M., Reisinger, H., Hofmann, F., Schulz, T., Landgraf, E., Luyken, R.J., Rösner, W., Grieb, M., Risch, L.: Charge trapping memory structures with Al_2O_3 trapping dielectric for high-temperature applications. Solid-State Electronics 49, 716–720 (2005)

[93] Sree Harsha, K.S.: Principles of Physical Vapor Deposition of Thin Films. Elsevier Ltd., Oxford (2006)

[94] Stolen, R., Dransfel, K.: Far-infrared lattice absorption in alkali halide crystals. Physical Review 139, 1295 (1965)

[95] Streiffer, S.K., Basceri, C., Parker, C.B., Lash, S.E., Kingon, A.I.: Ferroelectricity in thin films: The dielectric response of fiber-textured $(Ba_xSr_{1-x})Ti_{1+y}O_{3+z}$ thin films grown by chemical vapor deposition. Journal of Applied Physics 86, 4565–4575 (1999)

[96] Subbaswamy, K.R., Mills, D.L.: Theory of microwave-absorption in wide-band-gap insulators - the role of thermal phonon lifetimes. Physical Review B 33, 4213–4220 (1986)

[97] Sun, Y., Fidler, J.K.: Component value ranges of tunable impedance matching networks in rf communications systems. In: Seventh International Conference on HF Radio Systems and Techniques, pp. 185–189 (1997)

[98] Sutono, A., Heo, D., Emery Chen, Y.-J., Laskar, J.: High-q ltcc-based passive library for wireless system-on-package (sop) module development. IEEE Transactions on Microwave Theory and Techniques 49(10 Pt 1), 1715–1724 (2001)

[99] Sze, S.M., Ng, K.K.: Physics of semiconducotr devices, 3rd edn. John Wiley & Sons, Inc., New York (2006)

[100] Tagantsev, A.K.: Effect of a weak electric-field on the dielectric losses in centrally-symmetric ferroelectric substances of the displacement type. Zhurnal Eksperimentalnoi I Teoreticheskoi Fiziki 77, 1893–1904 (1979)

[101] Taylor, T.R., Hansen, P.J., Acikel, B., Pervez, N., York, R.A., Streiffer, S.K., Speck, J.S.: Impact of thermal strain on the dielectric constant of sputtered barium strontium titanate thin films. Applied Physics Letters 80, 1978–1980 (2002)

[102] Turalchuk, P., Vendik, I., Vendik, O., Berge, J.: Modelling of tune-able acoustic resonators based on BSTO films with induced piezo-electric effect. In: Proceedings of the 37th European Microwave Conference (2007)

[103] Vendik, O.G.: The damping of the ferroelectric modes in crystals of the $SrTiO_3$ type. Soviet Physics - Solid State 17, 1096 (1975)

[104] Vendik, O.G., Zubko, S.P.: Ferroelectric phase transition and maximum dielectric permittivity of displacement type ferroelectrics $(Ba_xSr_{1-x}TiO_3)$. Journal of Applied Physics 88, 5343–5350 (2000)

[105] Vendik, O.G., Zubko, S.P., Ter-Martirosayn, L.T.: Experimental evidence of the size effect in thin ferroelectric films. Applied Physics Letters 73, 37–39 (1998)

[106] Vorobiev, A., Gevorgian, S.: Tunable $Ba_xSr_{1-x}TiO_3$ FBARs based on SiO_2/W Bragg reflectors. In: 2010 IEEE MTT-S International Microwave Symposium Digest (MTT), pp. 1444–1447 (2010)

References

[107] Wang, X., Helmersson, U., Madsen, L.D., Ivanov, I.P., Munger, P., Rudner, S., Hjorvarsson, B., Sundgren, J.E.: Composition, structure, and dielectric tunability of epitaxial SrTiO$_3$ thin films grown by radio frequency magnetron sputtering. Journal of Vacuum Science & Technology A - Vacuum Surfaces and Films 17, 564–570 (1999)

[108] Waser, R. (ed.): Nanoelectronics and information technology, 2nd edn. Wiley-VCH, Weinheim (2005)

[109] Waterhouse, R., Shulley, N.: Full characterization of varactor loaded, probe-fed, rectangular, microstrip patch antennas. IEE Proc. Microw. Antennas Propag. 141(5), 367–373 (1994)

[110] Yildirim, N., Sen, O., Sen, Y.: Synthesis of cascaded n-tuplet filters. In: Proc. 5th International Conference on Telecommunications in Modern Satellite, Cable and Broadcasting Service, TELSIKS (2001)

[111] Yoon, Y., Kim, D., Allen, M., Kenney, J., Hunt, A.: A reduced intermodulation distortion tunable ferroelectric capacitor-architecture and demonstration. IEEE Transactions on Microwave Theory and Techniques 51(12), 2568–2576 (2003)

[112] Sun, Y., Fidler, J.K.: Design method for impedance matching networks. IEE Proceeding of Circuits, Devices and Systems 143(4), 186–194 (1996)

[113] Zhang, R., Yang, C., Yu, A., Wang, B., Tang, H., Chen, H., Zhang, J.: Wet chemical etching method for bst thin films annealed at high temperature. Applied Surface Science 254, 6697–6700 (2008)

[114] Zheng, Y., Giere, A., Jakoby, R.: A compact antenna with two independently tunable frequency bands. In: Antennas and Propagation Society International Symposium, AP-S. IEEE (2008)

[115] Zheng, Y., Hristov, A., Giere, A., Jakoby, R.: Suppression of harmonic radiation of tunable planar inverted-f antenna by ferroelectric varactor loading. In: IEEE MTT-S International Microwave Symposium Digest, pp. 959–962 (2008)

[116] Zhou, C., Newns, D.M.: Intrinsic dead layer effect and the performance of ferroelectric thin film capacitors. Journal of Applied Physics 82, 3081–3088 (1997)

Printed by Publishers' Graphics LLC
LSI20130404.15.35.78